冶金工业出版社
普通高等教育“十四五”规划教材

Materials Chemistry
材 料 化 学

Lei Xiping　Qiao Mingtao　Lei Wanying　Nan Yanli
雷西萍　乔明涛　雷琬莹　南艳丽　编著

Beijing
Metallurgical Industry Press
2023

Metallurgical Industry Press
39 Songzhuyuan North Alley, Dongcheng District, Beijing 100009, China

图书在版编目(CIP)数据

材料化学=Materials Chemistry:英文/雷西萍等编著.—北京:冶金工业出版社,2023.1

普通高等教育"十四五"规划教材

ISBN 978-7-5024-9229-8

Ⅰ.①材… Ⅱ.①雷… Ⅲ.①材料科学—应用化学—高等学校—教材—英文 Ⅳ.①TB3

中国版本图书馆 CIP 数据核字(2022)第 137583 号

Materials Chemistry
材料化学

出版发行	冶金工业出版社	**电　　话**	(010)64027926
地　　址	北京市东城区嵩祝院北巷 39 号	**邮　　编**	100009
网　　址	www. mip1953. com	**电子信箱**	service@ mip1953. com

责任编辑　杨　敏　美术编辑　彭子赫　版式设计　郑小利
责任校对　石　静　责任印制　窦　唯
北京印刷集团有限责任公司印刷
2023 年 1 月第 1 版,2023 年 1 月第 1 次印刷
787mm×1092mm　1/16;12.75 印张;307 千字;194 页
定价 39.00 元

投稿电话　(010)64027932　投稿信箱　tougao@cnmip. com. cn
营销中心电话　(010)64044283
冶金工业出版社天猫旗舰店　yjgycbs. tmall. com
(本书如有印装质量问题,本社营销中心负责退换)

Preface

Materials Chemistry is an interdisciplinary field of materials science and chemistry that covers the basics of materials science and physical chemistry, inorganic chemistry, and organic polymer chemistry. It is a discipline teaching how to use chemical knowledge to prepare materials and to understand the interaction between the nature of chemistry and the properties of materials. After careful study of the monographs and teaching materials of material chemistry on the market, it is found that there are many kinds of books about material chemistry, but each of them has its own emphasis and there is a lack of unity and comprehensiveness of the basic knowledge. At the same time, with the materials developing rapidly nowadays, new concepts are emerging as well. Material science researchers need to constantly learn and expand their knowledge of new materials, constantly improve thinking ability and cultivate innovative consciousness. The author also found one good textbook, *Materials Chemistry* (original English edition), published by Bradley D. Fahlman in 2012. However, its content consist of metal materials, solid state materials, polymer materials and materials characterization, does not start from the basic knowledge of chemistry, such as physical chemistry, inorganic chemistry, polymer chemistry, and even knowledge about materials, such as phase diagram, metastable state, amorphous state, are not covered. Therefore, the author thinks it is necessary to compile a more comprehensive English edition of materials chemistry, to introduce systematically the basic knowledge of chemistry on one hand, and materials science on the other hand, integrating the scientific research achievements of the editors, trying to make the concepts and theories more simple and easy to understand, and the structure of the knowledge system more perfect. At the end of each chapter, there are a few of questions to strengthen the understanding and utilization of the key knowledge. The publication of this textbook will not only pave the way for students of material specialty to go abroad for further study, but also provide a strong support for teaching the bilingual course of material chemistry.

Materials Chemistry is divided into eight chapters, Chapter 1 is *Foundation of Materials Science*, Chapter 2 is *Foundation of Physical Chemistry*, Chapter 3 is *Foundation of Polymer Materials*, Chapter 4 is *Preparation Principles in Materials Chemistry*, and Chapter 5 is *Preparation Methods of Metal*, Chapter 6 is *Preparation Methods of Inorganic Nonmetallic Material*, Chapter 7 is *Preparation Mechanisms of Polymer*, Chapter 8 is *Corrosion and Protection of Materials*. The first three chapters are the basic knowledge, the following four chapters are about the preparation mechanism and method, and the last chapter is about corrosion and protection of materials. The authors' scientific research results are also integrated into each chapter.

This textbook is compiled by four authors: Lei Xiping, Lei Wanying, Qiao Mingtao and Nan Yanli . Chapters 1 and 8. 1 ~ 8. 2 are edited by Nan Yanli, Chapter 2 and 8. 3 are edited by Lei Wanying, Chapter 3 and Chapter 7 are edited by Lei Xiping, Chapter 4, Chapter 5 and Chapter 6 are edited by Qiao Mingtao. The preface of the book is written by Lei Xiping. There will be mistakes, and we hope that readers do not hesitate to correct and make the book more perfect in use.

This textbook refers to many domestic and foreign monographs, textbooks, periodicals, reports and the academic achievements of the editors. Its publication will not be possible without the support of the Textbook Construction Foundation for Graduate Students of Xi'an University of Architecture and Technology (No. JC202009). We also acknowledge the professional guidance from Metallurgical Industry Press.

Xi'an University of Architecture and Technology
Lei Xiping
2022. 10

Contents

Chapter 1 Foundation of Materials Science

Sometimes it is useful to subdivide the discipline of materials science and engineering into materials science and materials engineering subdisciplines. Strictly speaking, materials science involves investigating the relationships that exist between the structures and properties of materials. In contrast, materials engineering involves, on the basis of these structure-property correlations, designing or engineering the structure of a material to produce a predetermined set of properties. From a functional perspective, the role of a materials scientist is to develop or synthesize new materials, whereas a materials engineer is called upon to create new products or systems using existing materials and/or to develop techniques for processing materials. Most graduates in materials programs are trained to be both materials scientists and materials engineers.

This chapter was concerned primarily with the various types of atomic bonding, which are determined by the electron structures of the individual atoms. The present discussion is devoted to the next level of the structure of materials, specifically, to some of the arrangements that may be assumed by atoms in the solid state. Within this framework, concepts of crystallinity and noncrystallinity are introduced. For crystalline solids, the notion of crystal structure is presented, specified in terms of a unit cell. The three common crystal structures found in metals are then detailed, along with the scheme by which crystallographic points, directions, and planes are expressed. Single crystals, polycrystalline materials, and noncrystalline materials are considered. Another section of this chapter briefly describes how crystal structures are determined experimentally using X-ray diffraction techniques.

1. 1 Crystal Structures

1. 1. 1 Unit Cells

The atomic order in crystalline solids indicates that small groups of atoms form a repetitive pattern. Thus, in describing crystal structures, it is often convenient to subdivide the structure into small repeat entities called unit cells. Unit cells for most crystal structures are parallelepipeds or prisms having three sets of parallel faces; one is drawn within the aggregate of spheres (Fig. 1. 1), which in this case happens to be a cube. A unit cell is chosen to represent the symmetry of the crystal structure, wherein all the atom positions in the crystal may be generated by translations of the unit cell integral distances along each of its edges. Thus, the unit cell is the basic structural unit or building block of the crystal structure and defines the crystal structure by virtue of its geometry and the atom positions within. Convenience usually dictates that parallelepiped corners coincide with centers of the hard-sphere atoms. Furthermore, more than a single unit cell may be chosen for a particular crystal structure; however, we generally use the unit cell having the highest level of geometrical symmetry.

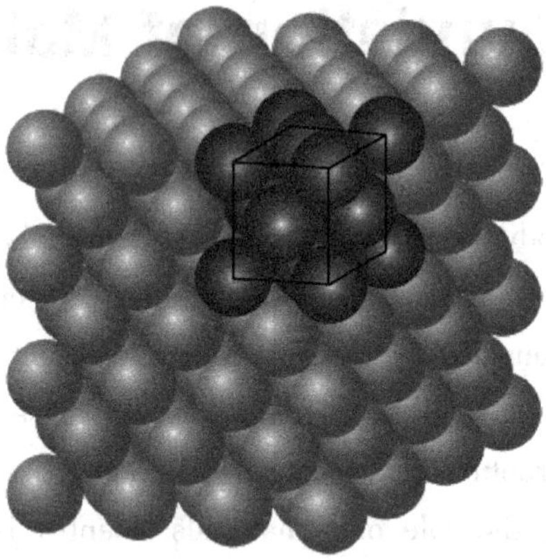

Fig. 1. 1 An aggregate of many atoms

1. 1. 2 Crystal Systems

Some metals, as well as non-metals, may have more than one crystal structure, a phenomenon known as polymorphism. When found in elemental solids, the condition is often termed allotropy. The prevailing crystal structure depends on both the temperature and the external pressure. One familiar example is found in carbon: graphite is the stable polymorph at ambient conditions, whereas diamond is formed at extremely high pressures. Also, pure iron has a BCC crystal structure at room-temperature, which changes to FCC iron at 912℃ (1674℉). Most often a modification of the density and other physical properties accompanies a polymorphic transformation.

Because there are many different possible crystal structures, it is sometimes convenient to divide them into groups according to unit cell configurations and/or atomic arrangements. One such scheme is based on the unit cell geometry, that is, the shape of the appropriate unit cell parallelepiped without regard to the atomic positions in the cell. Within this framework, xyz coordinate system is established with its origin at one of the unit cell corners; each of the x, y, and z axes coincides with one of the three parallelepiped edges that extend from this corner, as illustrated in Fig. 1. 2. The unit cell geometry is completely defined in terms of six parameters: the three edge lengths a, b, and c, and the three interaxial angles α, β, and γ. These are indicated in Fig. 1. 2, and are sometimes termed the lattice parameters of a crystal structure. On this basis there are seven different possible combinations of a, b, and c and α, β, and γ, each of which represents a distinct crystal system. These seven crystal systems are cubic, tetragonal, hexagonal, orthorhombic, rhombohedral, monoclinic, and triclinic. The lattice parameter relationships and unit cell sketches for each are represented in Table 1. 1. The cubic system, for which $a=b=c$ and $\alpha=\beta=\gamma=90°$, has the greatest degree of symmetry. The least symmetry is displayed by the triclinic system, because $a \neq b \neq c$ and $\alpha \neq \beta \neq \gamma$. From the discussion of metallic crystal structures, it should be apparent that both FCC and BCC structures belong to the cubic crystal system, whereas HCP falls within the hexagonal system. The conventional hexagonal unit cell really consists of three parallelepipeds situated as shown in Table 1.1.

What is the difference between crystal structure and crystal system? It is important to note that many of the principles and concepts addressed in previous discussions in this chapter also apply to crystalline ceramic and polymeric systems. For example, crystal structures are most often described in terms of unit cells, which are normally more complex than those for FCC, BCC, and HCP. In addition, for these other systems, we are often interested in determining atomic packing factors and densities, using modified forms of Eqs. (1.1) and (1.2). Furthermore, according to unit cell geometry, crystal structures of these other material types are grouped within the seven crystal systems.

$$\text{APM} = \frac{\text{volume of atoms in a unite cell}}{\text{total unite cell volume}} \tag{1.1}$$

Where the APF is the sum of the sphere volumes of all atoms within a unit cell (assuming the atomic hard-sphere model) divided by the unit cell volume.

$$\rho = \frac{nA}{V_C N_A} \tag{1.2}$$

where n is the number of atoms associated with each unit cell; A is the atomic weight; V_C is the volume of the unit cell; N_A is the Avogadro's number (6.022×10^{23} atoms/mol).

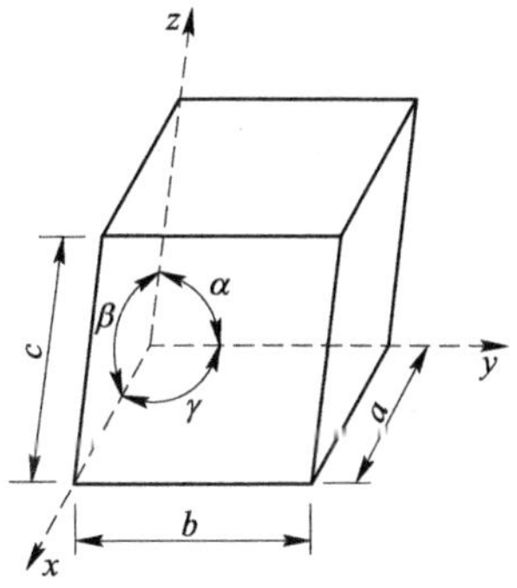

Fig. 1.2 A unit cell with x, y, and z coordinate axes, showing axial lengths (a, b and c) and interaxial angles (α, β and γ)

Table 1.1 Lattice parameter relationships and figures showing unit cell geometries for the seven crystal systems

Crystal system	Axial relationships	Interaxial angles	Unit cell geometry
Cubic	$a=b=c$	$\alpha=\beta=\gamma=90°$	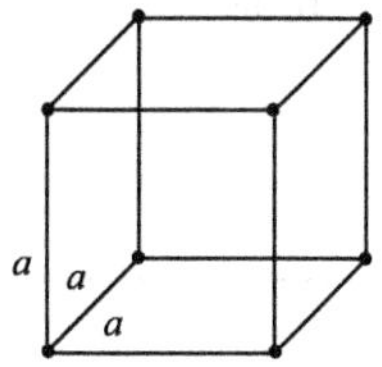

Continued Table 1. 1

Crystal system	Axial relationships	Interaxial angles	Unit cell geometry
Hexagonal	$a=b\neq c$	$\alpha=\beta=90°, \gamma=120°$	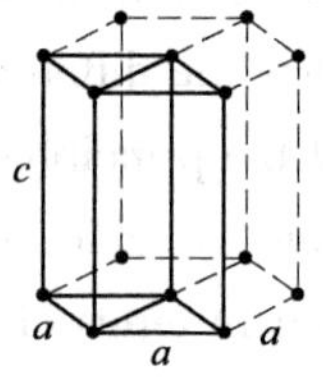
Tetragonal	$a=b\neq c$	$\alpha=\beta=\gamma=90°$	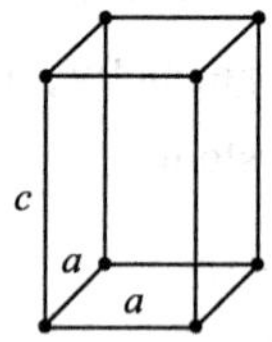
Rhombohedral (Trigonal)	$a=b=c$	$\alpha=\beta=\gamma\neq 90°$	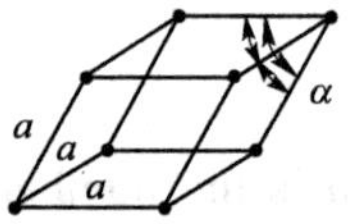
Orthorhombic	$a\neq b\neq c$	$\alpha=\beta=\gamma=90°$	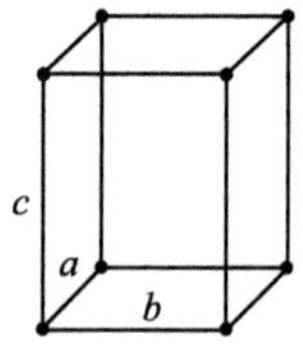
Monoclinic	$a\neq b\neq c$	$\alpha=\gamma=90°\neq\beta$	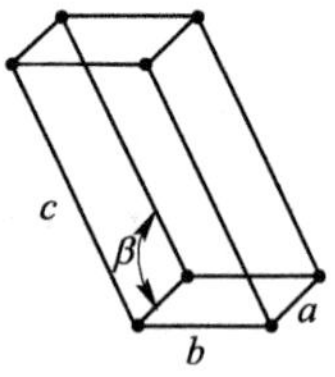
Triclinic	$a\neq b\neq c$	$\alpha\neq\beta\neq\gamma\neq 90°$	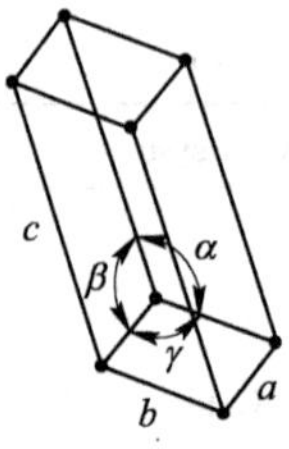

1. 2 Crystalline and Nanocrystalline Materials

1. 2. 1 Single Crystals

For a crystalline solid, when the periodic and repeated arrangement of atoms is perfector extends throughout the entirety of the specimen without interruption, the result is a single crystal. All unit cells interlock in the same way and have the same orientation. Single crystals exist in nature, but they can also be produced artificially. They are ordinarily difficult to grow because the environment must be carefully controlled.

If the extremities of a single crystal are permitted to grow without any external constraint, the crystal assumes a regular geometric shape having flat faces, as with some of the gemstones; the shape is indicative of the crystal structure. A garnet single crystal is shown in Fig. 1. 3. Within the past few years, single crystals have become extremely important in many modern technologies, in particular electronic microcircuits, which employ single crystals of silicon and other semiconductors.

Fig. 1. 3 A garnet single crystal that was found in Tongbei, Fujian province, China

1. 2. 2 Polycrystalline Materials

Most crystalline solids are composed of a collection of many small crystals or grains; such materials are termed polycrystalline. Various stages in the solidification of a polycrystalline specimen are represented schematically in Fig. 1. 4. Initially, small crystals or nuclei form at various positions. These have random crystallographic orientations, as indicated by the square grids. The small grains grow by the successive addition from the surrounding liquid of atoms to the structure of each. The extremities of adjacent grains impinge on one another as the solidification process approaches completion. As indicated in Fig. 1. 4, the crystallographic orientation varies from grain to grain. Also, there exists some atomic mismatch within the region where two grains meet; this area is called grain boundary

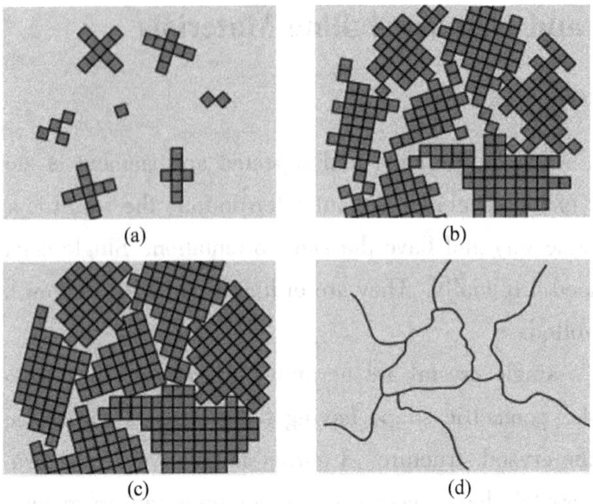

Fig. 1.4 Schematic diagrams of the various stages in the solidification of a polycrystalline material (the square grids depict unit cells)

(a) Small crystallite nuclei; (b) Growth of the crystallites (the obstruction of some grains that are adjacent to one another is also shown); (c) Upon completion of solidification, grains having irregular shapes have formed; (d) The grain structure as it would appear under the microscope (dark lines are the grain boundaries)

1.2.3 Anisotropy

The physical properties of single crystals of some substances depend on the crystallographic direction in which measurements are taken. For example, the elastic modulus, the electrical conductivity, and the index of refraction may have different values in the [100] and [111] directions. This directionality of properties is termed anisotropy, and it is associated with the variance of atomic or ionic spacing with crystallographic direction. Substances in which measured properties are independent of the direction of measurement are isotropic. The extent and magnitude of anisotropic effects in crystalline materials are functions of the symmetry of the crystal structure; the degree of anisotropy increases with decreasing structural symmetry-triclinic structures normally is highly anisotropic. The modulus of elasticity values at [100], [110], and [111] orientations for several metals are presented in Table 1.2. For many polycrystalline materials, the crystallographic orientations of the individual grains are totally random. Under these circumstances, even though each grain may be anisotropic, a specimen composed of the grain aggregate behaves isotropically. Also, the magnitude of a measured property represents some average of the directional values. Sometimes the grains in polycrystalline materials have a preferential crystallographic orientation, in which case the material is said to have a "texture". The magnetic properties of some iron alloys used in transformer cores are anisotropic—that is, grains (or single crystals) magnetize in a 100-type direction easier than any other crystallographic direction. Energy losses in transformer cores are minimized by utilizing polycrystalline sheets of these alloys into which have been introduced a *magnetic texture*: most of the grains in each sheet have a 100-type crystallographic direction that is aligned (or almost aligned) in the same direction, which is oriented parallel to the direction of the applied magnetic field.

Table 1.2 Modulus of elasticity values for several metals at various crystallographic orientations

Metal	Modulus of elasticity/GPa		
	[100]	[110]	[111]
Aluminum	63.7	72.6	76.1
Copper	66.7	130.3	191.1
Iron	125.0	210.5	272.7
Tungsten	384.6	384.6	384.6

1.2.4 Noncrystalline Solids

It has been mentioned that noncrystalline solids lack a systematic and regular arrangement of atoms over relatively large atomic distances. Sometimes such materials are also called amorphous (meaning literally "without form"), or supercooled liquids, inasmuchas their atomic structure resembles that of a liquid. An amorphous condition may be illustrated by comparison of the crystalline and noncrystalline structures of the ceramic compound silicon dioxide (SiO_2), which may exist in both states. Fig. 1.5(a) and (b) present two-dimensional schematic diagrams for both structures of SiO_2. Even though each silicon ion bonds to three oxygen ions for both states, beyond this, the structure is much more disordered and irregular for the noncrystalline structure. Whether a crystalline or an amorphous solid forms depends on the ease with which a random atomic structure in the liquid can transform to an ordered state during solidification. Amorphous materials, therefore, are characterized by atomic or molecular structures that are relatively complex and become ordered only with some difficulty.

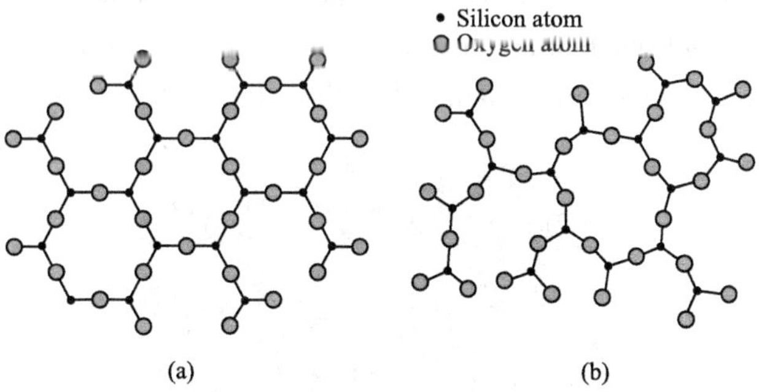

Fig. 1.5 Two-dimensional schemes of the structure
(a) Crystalline silicon dioxide; (b) Noncrystalline silicon dioxide

Furthermore, rapidly cooling through the freezing temperature favors the formation of a noncrystalline solid, because little time is allowed for the ordering process. Metals normally form crystalline solids, but some ceramic materials are crystalline, whereas others—the inorganic glasses—are amorphous. Polymers may be completely noncrystalline or semicrystalline consisting of varying degrees of crystallinity.

1.3 Defects

1.3.1 Point Defect

The simplest of the point defects is a vacancy, or vacant lattice site, one normally occupied but from which an atom is missing (Fig. 1.6). All crystalline solids contain vacancies, and, in fact, it is not possible to create such a material that is free of these defects. The necessity of the existence of vacancies is explained using principles of thermodynamics; in essence, the presence of vacancies increases the entropy (i. e., the randomness) of the crystal. The equilibrium number of vacancies N_v for a given quantity of material (usually per meter cubed) depends on and increases with temperature according to

$$N_v = N\left(-\frac{Q_v}{kT}\right) \tag{1.3}$$

In this expression, N is the total number of atomic sites (most commonly per cubic meter), Q_v is the energy required for the formation of a vacancy (J/mol or eV/atom), T is the absolute temperature in kelvins, and k is the gas or Boltzmann's constant. The value of k is 1.38×10^{-23} J/atom · K, or 8.62×10^{-5} eV/atom · K, depending on the units of Q_v. Thus, the number of vacancies increases exponentially with temperature—that is, as T in Eq. (1.3) increases, so also does the term $\exp(-Q_v/kT)$. For most metals, the fraction of vacancies N_v/N just below the melting temperature is on the order of 10^{-4}—that is, one lattice site out of 10000 will be empty. As ensuing discussions indicate, a number of other material parameters have an exponential dependence on temperature similar to that in Eq. (1.3).

A self-interstitial is an atom from the crystal that is crowded into an interstitial site—a small void space that under ordinary circumstances is not occupied. This kind of defect is also represented in Fig. 1.6. In metals, a self-interstitial introduces relatively large distortions in the surrounding lattice because the atom is substantially larger than the interstitial position in which it is situated.

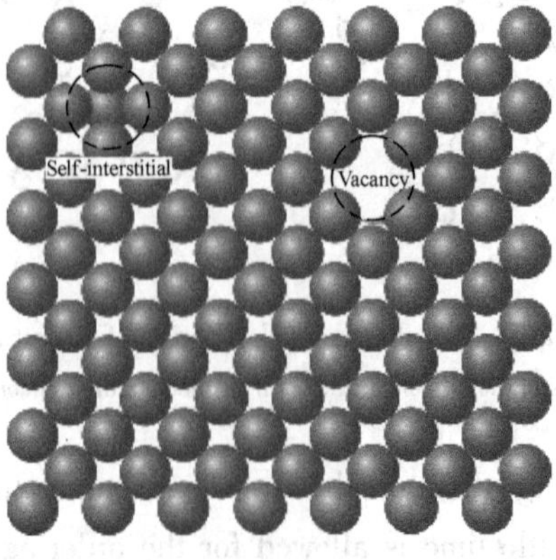

Fig. 1.6 Two-dimensional representations of a vacancy and a self-interstitial

Consequently, the formation of this defect is not highly probable, and it exists in very small concentrations that are significantly lower than for vacancies.

1.3.2 Linear Defects

A dislocation is a linear or one-dimensional defect around which some of the atoms are misaligned. One type of dislocation is represented in Fig. 1.7: an extra portion of a plane of atoms, or half-plane, the edge of which terminates within the crystal. This is termed an edge dislocation; it is a linear defect that centers on the line that is defined along the end of the extra half-plane of atoms. This is sometimes termed the dislocation line, which, for the edge dislocation in Fig. 1.7, is perpendicular to the plane of the page. Within the region around the dislocation line there is some localized lattice distortion. The atoms above the dislocation line in Fig. 1.7 are squeezed together, and those below are pulled apart; this is reflected in the slight curvature for the vertical planes of atoms as they bend around this extra half-plane. The magnitude of this distortion decreases with distance away from the dislocation line; at positions far removed, the crystal lattice is virtually perfect. Sometimes the edge dislocation in Fig. 1.7 is represented by the symbol ⊥, which also indicates the position of the dislocation line. An edge dislocation may also be formed by an extra half-plane of atoms that is included in the bottom portion of the crystal; its designation is a ⊤.

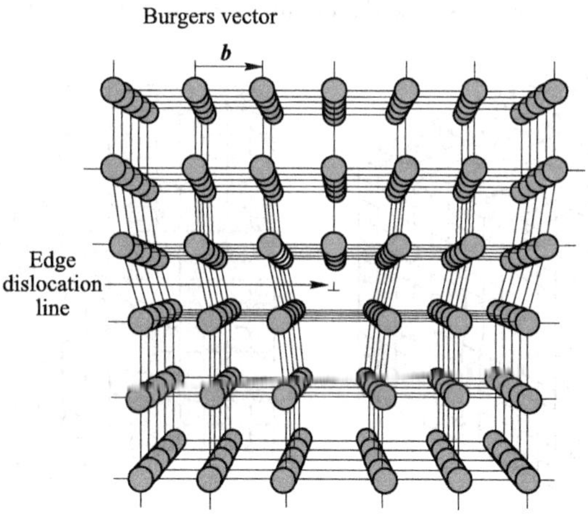

Fig. 1.7 The atom positions around an edge dislocation; extra half-plane of atoms

Another type of dislocation, called a screw dislocation, may be thought of as being formed by a shear stress that is applied to produce the distortion shown in Fig. 1.8(a): the upper front region of the crystal is shifted one atomic distance to the right relative to the bottom portion. The atomic distortion associated with a screw dislocation is also linear and along a dislocation line, line *AB* in Fig. 1.8(b). The screw dislocation derives its name from the spiral or helical path or ramp that is traced around the dislocation line by the atomic planes of atoms. Sometimes the symbol is used to designate a screw dislocation.

Most dislocations found in crystalline materials are probably neither pure edge dislocation nor pure screw dislocation but exhibit components of both types; these are termed mixed dislocations.

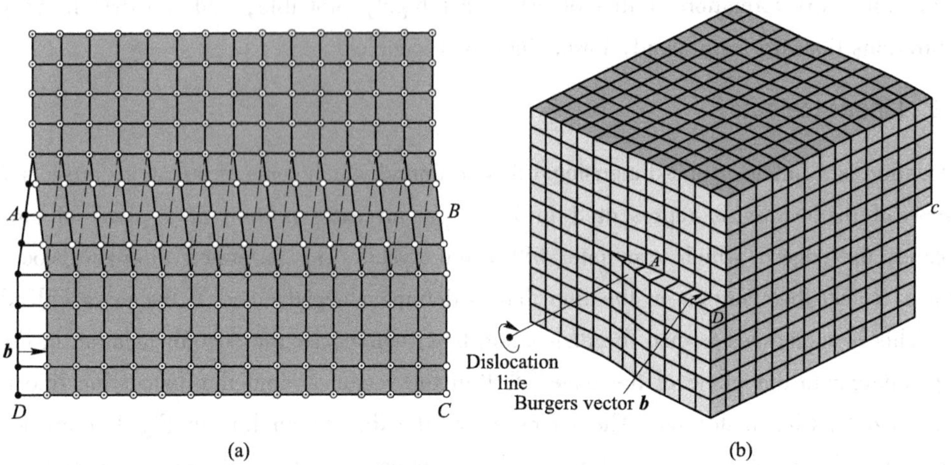

Fig. 1.8 Screw dislocation

(a) A screw dislocation within a crystal; (b) The screw dislocation in (a) as viewed from above (The dislocation line extends along line *AB*. Atom positions above the slip plane are designated by open circles, those below by solid circles)

All three dislocation types are represented schematically in Fig. 1.9; the lattice distortion that is produced away from the two faces is mixed, having varying degrees of screw and edge character.

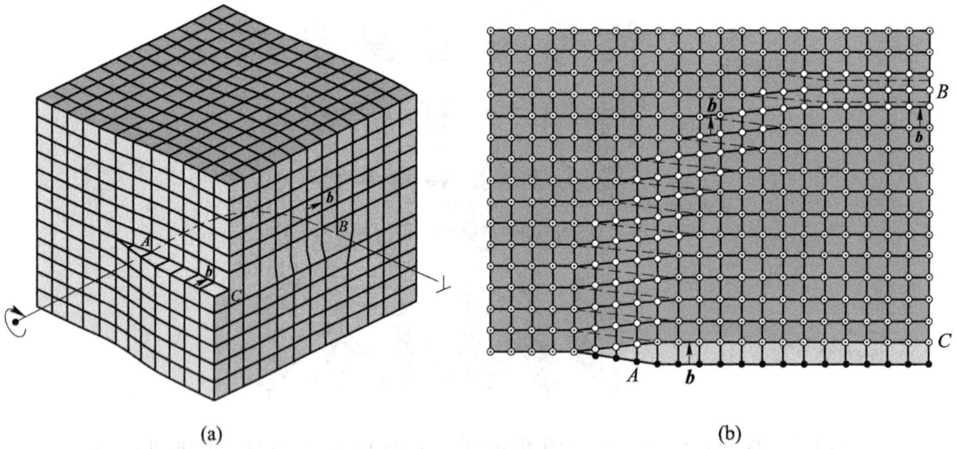

Fig. 1.9 Mixed dislocations

(a) Schematic representation of a dislocation that has edge, screw, and mixed character; (b) Top view, where open circles denote atom positions above the slip plane, and solid circles, atom positions below (At point *A*, the dislocation is pure screw, while at point *B*, it is pure edge. For regions in between where there is curvature in the dislocation line, the character is mixed edge and screw)

1.3.3 Interfacial Defects

Interfacial defects are boundaries that have two dimensions and normally separate regions of the materials that have different crystal structures and/or crystallographic orientations. These imperfections include external surfaces, grain boundaries, phase boundaries, twin boundaries, and stacking faults.

1.3.3.1 External Surfaces

One of the most obvious boundaries is the external surface, along which the crystal structure terminates. Surface atoms are not bonded to the maximum number of nearest neighbors and are therefore in a higher energy state than the atoms at interior positions. The bonds of these surface atoms that are not satisfied give rise to a surface energy, expressed in units of energy per unit area (J/m^2 or erg/cm^2). To reduce this energy, materials tend to minimize, if at all possible, the total surface area. For example, liquids assume a shape having a minimum area—the droplets become spherical. Of course, this is not possible with solids, which are mechanically rigid.

1.3.3.2 Grain Boundaries

Another interfacial defect, the grain boundary, was introduced in section 3.14 as the boundary separating two small grains or crystals having different crystallographic orientations in polycrystal-line materials. A grain boundary is represented schematically from an atomic perspective. Within the boundary region, which is probably just several atom distances wide, there is some atomic mismatch in a transition from the crystalline orientation of one grain to that of an adjacent one. Various degrees of crystallographic misalignment between adjacent grains are possible. When this orientation mismatch is slight, on the order of a few degrees, then the term *small-(or low-) angle grain boundary* is used. These boundaries can be described in terms of dislocation arrays. This type is called a *tilt boundary*; the angle of misorientation, u, is also indicated in the figure. When the angle of misorientation is parallel to the boundary, a *twist boundary* results, which can be described by an array of screw dislocations.

The atoms are bonded less regularly along a grain boundary (e.g., bond angles are longer), and consequently there is an interfacial or grain boundary energy similar to the surface energy just described. The magnitude of this energy is a function of the degree of misorientation, being larger for high-angle boundaries. Grain boundaries are more chemically reactive than the grains themselves as a consequence of this boundary energy. Furthermore, impurity atoms often preferentially segregate along these boundaries because of their higherenergy state. The total interfacial energy is lower in large or coarse-grained materials than in fine-grained ones because there is less total boundary area in the former. Grains grow at elevated temperatures to reduce the total boundary energy. In spite of this disordered arrangement of atoms and lack of regular bonding along grain boundaries, a polycrystalline material is still very strong; cohesive forces within and across the boundary are present. Furthermore, the density of a polycrystalline specimen is virtually identical to that of a single crystal of the same material.

1.3.3.3 Phase Boundaries

Phase boundaries exist in multiphase materials, in which a different phase exists on each side of the boundary; furthermore, each of the constituent phases has its own distinctive physical and/or chemical characteristics. As we shall see in subsequent chapters, phase boundaries play an important role in determining the mechanical characteristics of some multiphase metal alloys.

Questions

1. What is the difference between crystal structure and crystal system?
2. Do noncrystalline materials display the phenomenon of allotropy (orpolymorphism)? Why or why not?
3. The Fig. 1. 10 shows a unit cell for ahypothetical metal.
(a) To which crystal system does this unit cellbelong?
(b) What would this crystal structure be called?
(c) Calculate the density of the material, giventhat its atomic weight is 141g/mol.

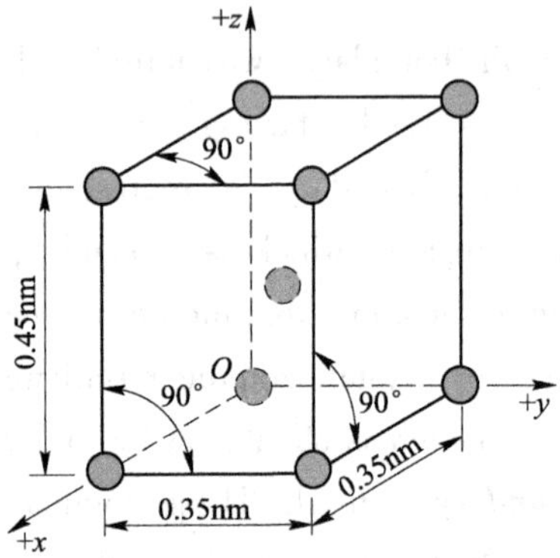

Fig. 1. 10 The unit cell for a hypothetical metal

4. Explain why the properties of polycrystalline materials are most often isotropic.

References

[1] Burns G, Glazer A M. Crystal systems-science direct[J]. Space Groups for Solid State Scientists (Third Edition), 2013, 76(2):25~44.
[2] Chow E. On the stress-strain relations for all crystal systems-science direct[J]. International Journal of Engineering Science, 1969, 7(3):247~256.
[3] Schomaker V, Lingafelter E C. Crystal systems[J]. Journal of Chemical Education, 1985, 62(3):219.
[4] Bartlelink D J, Biegesen D K, Johnson N M, et al. Single crystals: US, US4330363 A[P]. 1980.
[5] Swygenhoven H V. Polycrystalline materials. Grain boundaries and dislocations [J]. Science, 2002, 296(5565):66~67.
[6] Rosenhain W. An introduction to the study of physical metallurgy[M]. 2nd edition. Constable & Company Ltd., 1915.
[7] Hay J C, Sun E Y, et al. Elastic anisotropy of beta-silicon nitride whiskers[J]. Journal of the American Ceramic Society, 1998, 81(10):2661~2669.
[8] Cazacu O, Barlat F. A criterion for description of anisotropy and yield differential effects in pressure-insensitive metals[J]. International Journal of Plasticity, 2004, 20(11):2027~2045.

[9] Hertzberg R W. Deformation and fracture mechanics of engineering materials[M]. 3rd edition. John Wiley & Sons, 1989.

[10] Brailsford A D, Bullough R, Hayns M R. Point defect sink strengths and void-swelling[J]. Journal of Nuclear Materials, 1976, 60(3):246~256.

[11] Rizk R, Vajda P, et al. On the point defect production in electron-irradiated molybdenum[J]. Physica Status Solidi (a), 1972, 14(1):135~140.

[12] Moffatt W G, Pearsall G W, Wulff J. The structure and properties of materials[M]. Vol. Ⅰ, Structure, John Wiley & Sons, 1964.

[13] Chou Y T. Linear dislocation arrays in heterogeneous materials-science direct[J]. Acta Metallurgica, 1965, 13(7):779~783.

[14] Guy A G. Essentials of materials science[M]. McGraw-Hill Book Company, 1976.

[15] Read W T. Dislocations in crystals[M]. McGraw-Hill Book Company, 1953.

[16] Clark W, Pond R C. Interfacial defects in ordered and faulted materials[J]. Scripta Metallurgica, 1986, 20(8):1177~1181.

[17] Pond R C. Intrfacial defectsand epitaxy[C]. Mrs Online Proceedings Library Archive, 1985, 56.

Chapter 2 Foundation of Physical Chemistry

Physical chemistry is the study of macroscopic, atomic, subatomic and particulate phenomena in chemical systems in terms of the principles, practices and concepts of physics such as motion, energy, force, time, thermodynamics, statistical mechanics, analytical dynamics and chemical equilibrium. Physical chemistry is the study of the connection between physical and chemical phenomena of system based on the theory of the mathematics and physics as well as the experiment method. This chapter mainly introduces the foundation of physical chemistry including the following aspects: thermodynamics, chemical kinetics, and surface and interface chemistry, material electrochemistry.

2.1 Thermodynamic Laws

Based on the thermodynamic laws, chemistry thermodynamics is to solve the state change, phase change, and the direction of chemical reaction, etc. Also, thermodynamics is the study of energy and how it is converted form one form to another. Thermodynamics is divided into two subjects: classical thermodynamics and statistical thermodynamics. This section focuses on the classical thermodynamics. Classical thermodynamics is the study of the macroscopic properties of the system and the relationship between these properties. The laws of thermodynamics is provides the fundamental rules for the physics of the relationships between the variable of the system. Thermodynamics provides a powerful way to discuss equilibria and the direction of natural change in chemistry. Its concepts apply to both physical change, such as fusion and vaporization, and chemical change, including electrochemistry. We see that through the concepts of energy, enthalpy, entropy, Gibbs energy, and the chemical potential it is possible to obtain a unified view of these core features of chemistry and to treat equilibria quantitatively.

So far in this discussion of thermodynamics we have been talking about systems without having a definitive idea of what we exactly mean by a thermodynamic system. When we say thermodynamic system we refer to a macroscopic region of the universe that we define and select to analyze. This system includes the entire universe. Also, we choose the systems to study which are large enough that we can ignore the detailed structure of the material and focus on the system properties like its volume, pressure and temperature change. And the system may be affected by the surroundings. The systems could be classified to closed and open systems.

Closed systems do not exchange matter with their surroundings and have impermeable walls, while ***open systems*** do transfer matter across their boundaries and have permeable walls.

Energy, energy is the property of a system that can be converted into work. Energy is essentially an abstract concept in that it cannot be measured directly, but is determined by measuring other variables and coordinates of the system such as height, temperature and velocity.

2.1.1 The First Law of Thermodynamics

The first law of thermodynamics is the thermodynamic expression of one of the most basic laws of physics, the conservation of energy. This law could be simply stated that "energy cannot be created or destroyed" or that "the energy of the universe is a constant". The more precise statement is "A given amount of energy in a particular form can be converted to energy of a different form and then transformed back into the same amount of the original form, the total energy during the conversion and reverse process being constant".

Thermodynamics is the field which studies the transformation of energy. It is easy to understand that the rule which states that all energy transformations must leave the energy of the universe a constant is going to play a central role in thermodynamics. In this course the transformations of energy we are most interested in are those which involve materials and our main interest in the first law is how it governs the interaction of energy with materials. Since we know that as a material changes structure, or as the individual atoms of the material increase their motion the energy of the material may change. However it changes, we know that the change in energy of the material must be balanced by an equal and opposite change in energy of the surroundings. So although we haven't developed much of the detail of how energy and materials interact, we do know that whatever the details of their interaction the total energy is a constant throughout the process. Let's call the quantity which refers to the energy of our systems as U. U is the internal energy of the system, it refers to that energy associated with the internal condition of the system, and not on its spatial location or motion.

The internal energy U, is the energy associated with the internal condition of the systems.

The first law of thermodynamics, it represents the energy of a system; it is a function of the system variables of state; it changes between two states an amount which is independent of the process by which system was taken; the change in U of our system is balanced by an equal and opposite change of U of the surroundings such that there is no change in U for the universe no matter what process we perform.

So each system has an energy which depends on the condition of the system, but not on the process by which it arrived at the state. The change in this energy is balanced by a transfer of energy to or from the surroundings.

If we have a closed system, then the change in U is the heat flow from the surroundings plus the work done on the system by the surroundings. In other words, the only way to change the internal energy of our system is to allow energy to flow between it and its surroundings through its boundaries. This consequence of the first law is often written as

$$\Delta U = q + w \tag{2.1}$$

In fact, this equation is often used as a statement of the first law, while it is really just an accounting of how energy flowed into and out of a system during certain process.

If a system is closed, then we can choose the variables of state to be any two of the set of temperature T, pressure P and volume V. Here we choose variables of state to be P and V. The energy function is then

$$U = U(P, V) \tag{2.2}$$

and a change in U that arising from a process which changes P or V is written as

$$\Delta U = U_2 - U_1 = U(P_2, V_2) - U(P_1, V_1) \tag{2.3}$$

Where the system has been changed from state 1 (P_1, V_1) to state 2 (P_2, V_2). Note that this is a consequence of U being a state function, not the first law. Now in differential form ΔU becomes dU which is the total differential of U.

The expression $C_1(X_1, X_2, \cdots, X_n)\mathrm{d}X_1 + C_2(X_1, X_2, \cdots, X_n)\mathrm{d}X_2 + \cdots$ is an exact differential when it is equal to the total differential of a function written here as

$$\mathrm{d}Y = \left(\frac{\partial y}{\partial X_1}\right)\mathrm{d}X_1 + \left(\frac{\partial y}{\partial X_2}\right)\mathrm{d}X_2 + \cdots + \left(\frac{\partial y}{\partial X_n}\right)\mathrm{d}X_n = \sum_{i=1}^{n}\left(\frac{\partial Y}{\partial X_i}\right)_{X_j}\mathrm{d}X_i \tag{2.4}$$

where the X_i are independent variables. The necessary conditions for exactness are then that the functions C_i be equal to their corresponding partial derivatives of Y; $C_i = \left(\frac{\partial Y}{\partial X_i}\right)_{X_j}$ where the C_i and X_i are said to be conjugate to each other. The differential is then exact and $\mathrm{d}Y = \sum C_i \mathrm{d}X_i$.

In the case of the total differential of U mentioned above with the independent variables P and V:

$$\mathrm{d}U = \left(\frac{\partial U}{\partial V}\right)_P \mathrm{d}V + \left(\frac{\partial U}{\partial V}\right)_V \mathrm{d}T \tag{2.5}$$

Thus the consequence of the first law which is often used as an alternative expression of the first law; that the change in internal energy of a system during a process is equal to the sum of q, the heat that flows into the system from the surroundings and w the work done on the system by the surroundings during the process

$$\Delta U = q + w + w' + w'' + \cdots \tag{2.6}$$

where we divided the work into its different forms; w refers to the mechanical work done on the system. In differential form its expression becomes

$$\mathrm{d}U = \delta q + \delta w \tag{2.7}$$

where the differential δ denotes the fact that δq and δw are inexact differentials. Why are they not exact? Because they are not state functions. Notice that although δq and δw are inexact differentials their sum is an exact differential dU. In other words, although q and w are not state functions, their sum ΔU is. This alternative way of stating the conservation of energy, that $\Delta U = q + w$ will be our working statement of the first law.

We've already alluded to the fact that work plays an important role in thermodynamics. This is because energy, our central concept is itself defined as the potential to do work and work in thermodynamics represents a transfer of energy between the system and its surroundings. If this transfer of energy occurs due to some driving force other than a temperature gradient, then the energy transfer is due to work. Energy transfers due to gradients in temperature are heat flows. Now for flows of energy to a system other than heat the transfer can happen by the surroundings doing various different kinds of work on the system. Mechanical work can be done on the system, say by compressing the system (if the system boundary is movable). Work is the transfer of energy that changes the motion of atoms in the system in a uniform way. An example of electrical work being done on the system is moving charges in the system by the application of an external electric field. Generally, the work done is determined by the product of the generalized force projected onto a generalized displacement. Since we ignore the other work, and only consider the mechanical work in a system $\delta W = -\vec{F} \cdot \mathrm{d}\vec{l}$. Mechanical work done on the system due to the compression increases the internal energy of the system, whereas am expansion is associated with a decrease in the internal energy. The work to compress a gas or fluid at pressure P an infinitesimal amount $\mathrm{d}\vec{l}$ is

$$\delta W = -\vec{F} \cdot \mathrm{d}\vec{l} = -PA\mathrm{d}l = -P\mathrm{d}V \tag{2.8}$$

where A is the area over which the force is acting. The mechanical work is done on a system when there is a pressure gradient, that is pressure is the potential to do mechanical work. If the compression is done on a system which has adiabatic walls, that is where heat cannot be changed with the surroundings when $\delta q = 0$ and so $\mathrm{d}U = -P\mathrm{d}V$.

Like work, heat is a form of energy exchange between a system and its surroundings. The energy exchange between two objects by heat transfer occurs only when there is a temperature difference, or gradient between those two systems. Heat flows from regions of high temperatures to regions of low temperature. So like pressure, temperature is a potential for transferring energy, specifically the potential to transfer energy as heat. If the transfer occurs by conduction, then the rate of heat transfer is proportional to both the difference in temperature ΔT and the thermal conductivity k. From an atomic point of view, heat is the transfer of energy that occurs through the chaotic motion of matter at the molecular scale. The atoms in a hot region of a material vibrate chaotically more than atoms in a cooler region of a material. As atoms vibrate, they impart a force to their neighbors and cause them to move. The hotter the atoms, the more vigorous their motion and the larger the forces they impose on their neighboring atoms. This random motion that is passed from one point to another in the material results in energy transfer and eventually leads to a uniform amount of chaotic motion once the random motion of energetic atoms has flowed so that no temperature gradients persist. This transfer of kinetic energy to neighboring atoms accomplished through flow of random atomic motion is called heat transfer. So random, chaotic motion is disordered and classified as thermal motion, whereas work causes ordered, organized motion of the atoms in a system in a uniform manner.

Constant volume heat capacity, is the differential increase in internal energy of a system caused by a differential increase in temperature of the system with the volume held constant. The constant volume heat capacity is written as

$$C_V = \left(\frac{\partial U}{\partial T}\right)_V = \left(\frac{\partial q}{\partial T}\right)_V \tag{2.9}$$

Constant pressure heat capacity, is the differential heat flow into a system caused by a differential increase in temperature of the system held at constant pressure. The constant pressure heat capacity is written as

$$C_P = \left(\frac{\partial q}{\partial T}\right)_P \tag{2.10}$$

Although the relationship between the change in internal energy and the heat flow into the system for a constant volume system is simple and convenient, many processes occur at constant pressure, rather than at constant volume. This is because in many laboratory experiments our system is exposed to atmospheric pressure, rather than kept in a constant volume vessel. If the process is isobaric then the heat associated with this process is

$$q_P = (\Delta U + P\Delta V)_P \tag{2.11}$$

the quantity $\Delta U+P\Delta V$ shows up often in thermodynamics and for convenience we'll define it as the enthalpy.

Enthalpy, is the heat associated with a constant pressure process and is expressed as

$$H = U + PV \quad \text{and} \quad \Delta H = \Delta U + P\Delta V \tag{2.12}$$

Previously we saw that the heat associated with an isometric process was just equal to the change in internal energy. If volume is not held constant, the equality between the heat and the internal energy no longer holds because some of the energy supplied as heat to the system is returned to the surroundings as work, and thus $\mathrm{d}U < \delta q$ since $\delta w < 0$. Although the heat supplied to the system is no longer equal to the internal energy of the system, it is equal to another thermodynamic property of the system, the *enthalpy* H. Note from the above definition that a) enthalpy has the units of energy, and b) thatenthalpy is a state function since U, P and V all are.

$$H + \mathrm{d}H = U + \mathrm{d}U + (P + \mathrm{d}P)(V + \mathrm{d}V) \tag{2.13}$$

Since $\mathrm{d}P=0$,

$$H + \mathrm{d}H = U + \mathrm{d}U + P(V + \mathrm{d}V) \tag{2.14}$$

and using the definition $H = U + PV$,

$$\mathrm{d}H = \mathrm{d}U + P\mathrm{d}V \tag{2.15}$$

From the first law and considering only mechanical work, $\mathrm{d}U = \delta q_P - P\mathrm{d}V$,

$$\mathrm{d}H = \delta q_P - P\mathrm{d}V + P\mathrm{d}V \tag{2.16}$$

and finally that,

$$\mathrm{d}H = \delta q_P \tag{2.17}$$

Now if we go back to our constant volume calorimeter and take off its top so that our process occurs at the laboratory pressure we can equate the heat we supply to the *enthalpy* system to the change of our system for that process.

If the system that we are studying happens to have a very small volume change during the constant pressure process, and if the pressure isn't too large (these are the conditions $P\mathrm{d}V$ for being small) we see that

$$\mathrm{d}H = \mathrm{d}U + P\mathrm{d}V \approx \mathrm{d}U \tag{2.18}$$

This is exactly the case we find for processes involving condensed phases for low to moderate pressures. In the case of condensed phases at moderate pressures the change in volume during a process does very little work against the atmospheric pressure and so the change in internal energy of the system is approximately equal to the change in enthalpy of the system.

2.1.2 The Second Law and the Third Law of Thermodynamics

Consider the following system. A ball is dropped from a height *d*. From experience we know that with each successive bounce the ball will reach a height lower than its previous height. Microscopically, we'd see that each impact converts some of the uniform motion of the falling ball into chaotic motion of the atoms in the ball and the ground and consequently that the temperature of the ball and the ground increases. This process is said to *thermalize* the energy of the collision such that a portion of the work done on the ball and ground during impact is converted into heat. In our experience we encounter no circumstances where the disorganized motion of the atoms in the ball or ground that we associate with the thermal energy of the two bodies is completely converted back into the organized motion of macroscopic motion of the ball. For example, balls to not bounce higher than their original height. Here we see that there is a natural direction for the process of a ball bouncing from a given height; the ball bounces lower with each successive bounce and the work done on the ball and ground is converted into heat.

Spontaneous process is one which proceeds without an external driving force. All processes occurring in nature without the interference of an externally directed force are spontaneous.

Heat naturally flows from a region of high temperature to a region of low temperature. No external force is needed to move heat down a temperature gradient. Likewise, a drop of dye placed on the surface of water naturally mixes with the water without any stirring from an external force. If the individual molecules in the dye move randomly with no bias in any direction the drop still disperses and the process leads to mixing with no external driving force causing the molecules in the dye to move away from each other. Similarly, gas molecules spontaneously move from regions of high pressure to regions of low pressure. The pressure gradient itself drives the mass transfer with no external driving force.

The reverse of these processes do not occur without some external action. Unmixing the dye from the water is not a spontaneous process. The dye molecules once mixed into the water are dispersed and purifying the water or dye is an effort against *entropy*. If the dye spontaneously separated itself from the water into a small drop it would mean that the randomness (or entropy) of the unmixed system is higher than the randomness (entropy) of the mixed system, which is not the case (assuming that the dye and water molecules are not immiscible). Sometimes processes appear to spontaneously move towards a decrease in randomness (the unmixing of oil and water, for example), although closer inspection reveals that this is not the case. We'll discuss this paradox in more detail later. Notice that all these processes have a natural direction. If we video taped these processes and showed them to you in reverse you would immediately know that the tape was being played backwards. In other words, these processes are not reversible.

A reversible process is one which reverses itself with an infinitesimal driving force. No infinitesimal driving force is going to unmix the dye from the water, nor will a minute change in the conditions of my system cause heat to flow back up a temperature gradient it just flowed down. Since the driving forces in the spontaneous processes described above are all finite, an infinitesimal driving force to oppose the spontaneous process cannot overcome the natural progression of the process. The concept of reversibility here is somewhat abstract since it is only in the ideal sense that we could ever apply an infinitesimal driving force. For example, if the temperature gradient I create in my bar of iron is infinitesimally small, heat will flow very slowly from the higher temperature end to the lower temperature end. If I then apply an infinitesimal driving force to create a temperature gradient in the reverse direction, heat will flow in the reverse direction. That is, the original process of heat flowing down a temperature gradient is reversible if the original driving force is very small. Notice too that this process becomes very slow when it is reversible because the infinitesimal driving force causes the process to proceed very slowly. Using these principles we can then mentally test whether a process is spontaneous, or reversible by these two rules:

(1) If the process appears natural in reverse, then it is not spontaneous.

(2) If the process can be reversed by application of an infinitesimal external driving force, then it is reversible.

An additional handy mental test might be this:

If the process is proceeding at a finite rate it is not reversible.

The last section on spontaneous processes and reversibility was difficult to write without using the concept of entropy. However, I was able to write those sections without using the word entropy by using terms such as disorder, randomness, and chaotic. This illustrates just how useful the concept of entropy is since it allows us to develop a precise theory about randomness, reversibility and spontaneous processes. In fact we are going to use the concept of entropy as the foundation for stating the second law of thermodynamics. The natural way to proceed is to be more precise about what we mean by entropy.

What is the entropy? It is a measure of the disorder in the system. An increase in entropy of a system means that its randomness has increased. A solid crystal is more ordered than its liquid since the crystal atoms are arranged in a well defined pattern. The atoms of the liquid move about randomly. Likewise, the vaporization of the liquid results in a higher entropy phase because the motion and configuration of the atoms of the liquid were more ordered than the motions and configuration of the atoms in the gas. The entropy of the a drop of dye placed in a glass of water increases as the dye diffuses and the system mixes. As heat flows down a temperature gradient, entropy increases because energy is dispersed in the system more randomly.

Although all this is correct, we haven't said anything quantifiable yet or anything that can help us solve a thermodynamics problem. To do this we'll have to develop more of the connection between entropy, disorder, reversibility and spontaneous processes, and then we can get back to developing the quantitative rules associated with entropy.

Disorder is related to the degree to which energy, atomic configuration, and motion in a system are not ordered. For example, if the atoms in a solid all move together in a uniform way, there is no disorder in their motion. On the other hand, if each atom follows a trajectory which appears independent of the other atoms in the system then the motion of the atoms is disordered, and random. One way to see how much disorder is in a system is to analyze how much detail is required to specify the microscopic state of a system. An ordered system requires very little data to describe it. For example, the location of atoms in a perfect crystal can be specified with the lattice type, basis and lattice constants. On the other hand, the exact structure of a disordered alloy would be impossible to specify. We connect the concepts of disorder and spontaneity by noticing that the amount of order tends to naturally decrease as a process proceeds.

Spontaneity is naturally associated with processes which increase the disorder or entropy of the universe. The entropy of the ball and ground is higher on the nth bounce than it was on the $(n-1)^{th}$ bounce because some of the energy of the ball was converted into the disordered motion of the atoms in the ball and the ground. The entropy of the mixture of dye and water increases with time since at any given point in time our knowledge of the location of the dye decreases. The direction of processes is associated with disorder increasing with time. Furthermore, if processes do not lead to overall decreases in the amount of disorder in the universe, then a process cannot be reversed if it increased the disorder of the universe because in order to reverse it we would have to somehow create the original amount of order from the state of increased disorder.

Entropy is functionally related to the measurable coordinates which characterize the system and which is an intrinsic property of the system related to the degree of disorder within the system. For a reversible process, changes in the entropy are given by:

$$dS = \frac{\delta q}{T} \tag{2.19}$$

This relationship between the heat of the process, the temperature and the differential change in entropy of the process only holds for reversible processes. Note also that since the entropy is functionally related to the measurables that characterize the system it must be a state function.

The second law of thermodynamics, the change in entropy of the universe (that is the system and its surroundings considered together) is positive and approaches zero for any process which approaches reversibility.

$$\Delta S \geqslant 0 \tag{2.20}$$

In other words, a process increases the entropy (disorder) of the universe, unless it is reversible in which case it leaves the entropy of the universe unchanged. Remember, that a process is said to be reversible if its direction can be reversed at any given point in time by an infinitesimal change in the external conditions of the system. So if a process is reversible it must have not been creating entropy since once it has created entropy that entropy cannot be destroyed. This is again because if entropy has increased, then disorder has increased. To return to the state of order from a state of higher disorder would require creating order from disorder, which is impossible, as is implicitly stated by the second law. Notice then that the second law also implies that a process must be proceeding infinitesimally slowly to not be creating entropy. That is, if the driving forces for the process are finite, then an infinitesimal force cannot reverse them. Since the process cannot be reversed by the infinitesimal driving force then it must be creating entropy so a process which proceeds at a finite rate is not reversible.

For example, if I am melting ice by adding heat very slowly, I can reverse the melting, that is I can refreeze the water by removing heat slowly from the system. This process can be reversed because if it is done slowly it is kept very close to its equilibrium condition because the driving forces for the process are kept infinitesimally small. If I heat the ice too rapidly, then some of the heat might lead to causing the liquid water to heat up, which would cause the molecules to move quickly and dissipate their energy in ways other than in melting the ice. Another example is the isothermal expansion of a gas. Lowering the pressure on the gas slightly allows the gas to expand slowly. Increasing the pressure again causes the gas to compress back to its original state. If the pressure is lowered more than an infinitesimal amount then the gas will expand rapidly, creating tempera-ture and pressure gradients and heat flow and dissipation of the energy of the gas.

Another statement of the second law, which can either be viewed as a result of our statement of the 2nd law or as an alternative statement for it is: *No process is possible in which the sole result is the adsorption of heat from a reservoir and its complete conversion into work.*

Some heat is always discarded during the process to a cold sink which increases the entropy of the cold sink. This is again related to the fact that order cannot be created from disorder. Creating ordered motion (work) from disordered motion (heat) can be done, but the overall process produces entropy.

For a reversible process we noted that $dS = \frac{\delta q}{T}$ thus to determine the change in entropy of a system for a reversible process we just integrate dS,

$$\Delta S = S_2 - S_1 = \int_1^2 \frac{\delta q}{T} \tag{2.21}$$

The entropy change of the surroundings is just the negative of the entropy change of the system for a reversible process. This is because during a reversible process the only heat involved is the transfer of energy between the system and the surroundings via heat. No additional random motion is generated, either in the system or in the surroundings so this random motion is not transferred between the system and the surroundings through heat. Because entropy is an state function we know that the change in entropy for a process which returns the system to its original state (say a cyclic process) leaves the entropy of the system unchanged. The entropy of the surroundings will be left unchanged as well, as long as the cycle was performed reversibly.

The first law states the conservation of energy as $dU = \delta q + \delta w$, where as the second law states the reversible process $dS = \frac{\delta q}{T}$. If we only consider reversible processes and mechanical work $\delta w = -PdV$, we can combine these two statements into the expression:

$$dU = TdS - PdV \tag{2.22}$$

which is one of the most important equations in thermodynamics and we will use it extensively to derive relationships among the thermodynamic variables.

We now address the problem of the value of $S(0)$. At $T = 0$, all energy of thermal motion has been quenched, and in a perfect crystal all the atoms or ions are in regular, uniform array. The localization of matter and the absence of thermal motion suggest that such materials also have zero entropy. This conclusion is consistent with the molecular interpretation of entropy, because $S=0$ if there is only one way of arranging the molecules and only one microstate is accessible (all molecules occupy the ground state)

The third law of thermodynamics, the entropy of all perfect crystalline substances is zero at $T=0$.

2.1.3 Thermodynamic Variables and Relations

In order to simplify our derivations of thermodynamic variable relationships we will apply Legendre transformations to our combined statement of the first and second law as equation (2.19).

First, let's decide to make a new function out of U which replaces one of its natural variables, volume V with pressure P. We define a function H which subtracts the conjugate pair $(-PV)$ from the function U:

$$H \equiv U + PV \tag{2.23}$$

and so the differential of this new function is just

$$\mathrm{d}H = \mathrm{d}U + P\mathrm{d}V + V\mathrm{d}P \tag{2.24}$$

If we substitute in our combined statement of the first and second laws:

$$\mathrm{d}H = T\mathrm{d}S - P\mathrm{d}V + P\mathrm{d}V + V\mathrm{d}P \tag{2.25}$$

Thus,

$$\mathrm{d}H = T\mathrm{d}S + V\mathrm{d}P \tag{2.26}$$

which as we planned has replaced the variable of volume with its conjugate pair pressure P. That is, this new function H has natural variables of entropy S and pressure P, $H=H(S, P)$. If during a process we hold pressure constant that

$$\mathrm{d}HP = T\mathrm{d}S \tag{2.27}$$

and that if the process is carried out quasistatically, or reversibly in other words, that we can equate $T\mathrm{d}S$ to the heat flow so that

$$\mathrm{d}HP = \delta q_{\mathrm{rev}} \tag{2.28}$$

Another transformation we could have made is defining a new function to replace the entropy S with the temperature as the natural variable. Do to this we subtract the conjugate pair $T\mathrm{d}S$ from U which defines a new function F:

$$F = U - TS \tag{2.29}$$

with the natural variables of volume and temperature. Taking the total derivative of this we find that

$$\mathrm{d}F = \mathrm{d}U - T\mathrm{d}S - S\mathrm{d}T \tag{2.30}$$

and if we substitute in the combined statements of the first and second laws as we did in the case of enthalpy results in

$$\mathrm{d}F = T\mathrm{d}S - P\mathrm{d}V - T\mathrm{d}S - S\mathrm{d}T \tag{2.31}$$

Thus,

$$\mathrm{d}F =- S\mathrm{d}T - P\mathrm{d}V \tag{2.32}$$

where our newly defined function F has natural variables of T and V, $F=F(T, V)$, we call this new function the ***Helmholtz free energy***. Notice that if we have a process which is carried out reversibly at constant temperature that

$$dF_T = -PdV \tag{2.33}$$

This says that the Helmholtz free energy is then just equal to the work done on the system during an isothermal process.

$$dF_T = \delta w_{rev} \tag{2.34}$$

The last possible Legendre transformation we can perform on a system with only two independent variables is to replace both independent natural variables simultaneously. In the case of $U(S, V)$, we'll create a new function with temperature and pressure as the new natural variables. To do this let's just define the new function $G=G(T, P)$ with

$$G \equiv U - TS + P \tag{2.35}$$

Taking the total derivative of this we find that

$$dG = dU - TdS - SdT + PdV + VdP \tag{2.36}$$

Substituting in the expression for the combined first and second laws results in:

$$dG = TdS - PdV - TdS - SdT + PdV + VdP \tag{2.37}$$

which we simplify to obtain:

$$dG = -SdT + VdP \tag{2.38}$$

This function $G=G(T, P)$ would be the most practical energy function in many cases since it has the most convenient pair of natural variables, pressure and temperature which are both easy to control in the laboratory. The internal energy on the other hand, although it is theoretically convenient (because we can connect changes in U to heat flow and work done) has the inconvenient variables of S and V. Inconvenient because performing experiments in a constant volume chamber is more cumbersome than using a constant pressure condition, and because creating a constant entropy condition is not well understood, let alone practical.

Above we defined three new functions using a Legendre transformation on the internal energy. Note that the Legendre Transformation doesn't change the units. That is the units of the new function are the same as the units of the original function since they are both made up of conjugate pairs whose products have the same units. In this particular case, the three new functions have the units of energy, as does the internal energy of the system. We called the first "new" function H, the enthalpy. We had already met this thermodynamic quantity and you might have noticed that we used the same definition here as we used when introducing enthalpyearlier, that

$$H \equiv U + PV \tag{2.39}$$

2.2 Chemical Reaction Kinetics

2.2.1 Chemical Reaction and Rate Law

2.2.1.1 Reaction System

In order to make our discussion simpler, let us assume for the time being that our system is closed with respect to matter exchange and has a fixed volume. Otherwise, there would be concentration changes not only due to chemical reactions, but also due to mass exchange or diffusion across the boundary and volume change, respectively. Obviously, it is much easier to treat such a closed and constant volume system.

Secondly, we assume that our reaction system is held under isothermal conditions. This means that heat transfer is very much faster than the chemical reactions taking place inside, and thus our entire reaction system is kept at a constant temperature. As the other extreme, if the chemical reactions are much faster no matter whether exothermic or endothermic, then we may treat the reaction system as adiabatic.

Thirdly, we assume that our reactions are homogeneous with respect to the reaction sites. In reality, chemical reactions occur much more often at heterogeneous reaction sites, e.g., catalysts. For simplicity's sake here, we treat only homogeneous reactions taking place within the interior of the system.

2.2.1.2 Chemical Reaction Rate

Let us consider the chemical reaction,

$$aA + bB + \cdots = cC + dD + \cdots \tag{2.40a}$$

where A, B, ⋯ stand for the reacting species and a, b, ⋯ the stoichiometric coefficients. The reacting species on the left hand side are normally called the reactants and those on the right-hand side the products.

It is often convenient to write this chemical reaction as

$$0 = (-a)A + (-b)B + \cdots + cC + dD + \cdots = \sum_{S} v_S S \tag{2.40b}$$

where S denotes the reacting species A, B, ⋯, and v_S the stoichiometric coefficient of S. Then, it turns out that a stoichiometric coefficient does have an algebraic sign: $v_S < 0$ for the reactants and $v_S > 0$ for the products

The reaction rate of a chemical reaction Eq. (2.41), R, is usually defined as the decreasing rate of a reactant or the increasing rate of a product to make it formally (or officially) positive or, denoting the concentration of a reacting species S(¼A, B,⋯C,D,⋯) as c_S,

$$R = -\frac{dcA}{dt};\ -\frac{dcB}{dt};\cdots;\ -\frac{dcC}{dt};\ -\frac{dcD}{dt};\cdots \tag{2.41}$$

Then, the reaction rate of a given reaction should take different values depending on the choice of S, and hence, looks somewhat awkward. We rather like to have a common value for the given reaction, irrespective of the choice of S. You may notice that once divided by a reacting species' own stoichiometric coefficient, the rate R turns universal or

$$R = \frac{1}{v_S} \times \frac{dc_S}{dt} \equiv \frac{d\xi}{dt} \tag{2.42}$$

where ξ as such is called the extent or advancement of reaction. The rate of a chemical reaction is normally defined in this way.

2. 2. 1. 3 Concentration Change in General

As the reaction rate under a closed, constant-volume condition has now been appropriately defined, let us think about the concentration change in general, namely, when the boundary is left open with respect to material exchange as well as the volume is allowed to change. The concentration change of a species dc is then the sum of the changes due to chemical reaction (rxn), diffusion (diff), and volume change (vol) or

$$\mathrm{d}c = \left(\frac{\delta c}{\delta t}\right)_{\mathrm{rxn}} \mathrm{d}t + \left(\frac{\delta c}{\delta t}\right)_{\mathrm{diff}} \mathrm{d}t + \left(\frac{\delta c}{\delta V}\right)_{\mathrm{vol}} \mathrm{d}V \tag{2.43a}$$

or dividing through by dt,

$$\frac{\mathrm{d}c}{\mathrm{d}t} = \left(\frac{\delta c}{\delta t}\right)_{\mathrm{rxn}} + \left(\frac{\delta c}{\delta t}\right) + \left(\frac{\delta c}{\delta V}\right)_{\mathrm{vol}} \frac{\mathrm{d}V}{\mathrm{d}t} \tag{2.43b}$$

We have just learned that the concentration change by a chemical reaction is, due to Eq. (2.44),

$$\left(\frac{\delta c}{\delta t}\right)_{\mathrm{rxn}} = vR \tag{2.44}$$

The change by diffusion is written as

$$\left(\frac{\delta c}{\delta t}\right)_{\mathrm{diff}} = \nabla D \nabla c \tag{2.45}$$

where D is the appropriate diffusion coefficient.

As $c = n/V$ with n being the number of moles of the species in the system, its concentration change due to volume change is

$$\left(\frac{\delta c}{\delta V}\right)_{\mathrm{vol}} \frac{\mathrm{d}V}{\mathrm{d}t} = -\frac{c}{V}\frac{\mathrm{d}V}{\mathrm{d}t} \tag{2.46}$$

Thus, Eq. (2.47) finally takes the form in general,

$$\frac{\mathrm{d}c}{\mathrm{d}t} = vR + \nabla D \nabla c - \frac{c}{V}\frac{\mathrm{d}V}{\mathrm{d}t} \tag{2.47}$$

We will now be dealing with a constant-volume system with no diffusion.

2. 2. 1. 4 Reaction Rate Law

In general, the reaction rate R as defined in Eq. (2.48) is a function of the concentrations of all the species present in a reaction system or

$$R = f(c_1, c_2, \cdots, c_j, \cdots) \tag{2.48}$$

This sort of relationship is called ***the rate law*** for a given reaction. The functional form of the rate law is still to be determined by experiment and cannot be a priori predicted from the stoichiometric coefficients of the reacting species of a given reaction, Eq. (2.49). Generally, it has nothing to do with the stoichiometric coefficients and everything to do with the reaction mechanisms.

In many cases, though, it has been found that the rate law may be represented empirically as

$$R = kc_1^{m1} c_2^{m2} \cdots c_j^{mj} \cdots \tag{2.49}$$

where the exponent m_j is called ***the partial order of reaction*** or partial reaction order with respect to the component j, and $m(=\sum m_j)$, the overall order of reaction or overall reaction order. The partial reaction orders m_j are not necessarily integers and are normally determined empirically. Here, the proportionality constant, k is called the rate law constant or simply, rate constant. Because this corresponds to the rate R when $c_1 = c_2 = \cdots = c_j = \cdots = 1$ (for whatsoever unit of c_j), it is often called the specific reaction rate. It should be noted that the specific rate or rate constant, k has a unit which depends on the overall reaction order m of the given reaction.

2.2.2 The Integral form of the Reaction Rate Equation

2.2.2.1 Differential Rate Equation

For the reaction

$$v_A A \longrightarrow v_B B + v_C C$$

The rate law of the form

$$v = k \cdot c(A)^m$$

can be expressed as

$$-\frac{1}{v_A} \cdot \frac{dc(A)}{dt} = k \cdot c(A)^m \tag{2.50}$$

which constitutes ***the differential form of the rate equation***.

The exponent in the rate law, m, determines the order of the reaction; it is said to proceed with the m-th order. The differential rate equations for the different reaction orders are:

$$-\frac{1}{v_A} \cdot \frac{dc(A)}{dt} = k \cdot c(A)^0 = k \qquad 0^{th} \text{ order} \tag{2.51}$$

$$-\frac{1}{v_A} \cdot \frac{dc(A)}{dt} = k \cdot c(A)^1 = k \cdot c(A) \qquad 1^{st} \text{ order} \tag{2.52}$$

$$-\frac{1}{v_A} \cdot \frac{dc(A)}{dt} = k \cdot c(A)^2 \qquad 2^{nd} \text{ order} \tag{2.53}$$

$$-\frac{1}{v_A} \cdot \frac{dc(A)}{dt} = k \cdot c(A)^3 \qquad 3^{rd} \text{ order} \tag{2.54}$$

2.2.2.2 Integrated Rate Expression for the First Order Rate Law

In order to derive an expression for a given rate law that allows calculation of numerical values for the concentration of a reactant over time, the differential rate expression needs to be integrated:

$$v = -\frac{1}{v_A} \cdot \frac{dc(A)}{dt} \rightarrow c_t(A) = ?$$

To integrate ***the differential 1st-order rate law***

$$-\frac{1}{v_A} \cdot \frac{dc(A)}{dt} = k \cdot c(A) \tag{2.55}$$

One first identifies the two variables that are of interest for the integration; here, these are $c(A)$ and t. The equation is then re-arranged such that the two variables get isolated on opposite sides:

$$-\frac{1}{v_A} \cdot \frac{1}{c(A)} \cdot dc(A) = k \cdot dt$$

The integration has to be carried out with respect to the two differential variables, i. e. $\frac{1}{c(A)} \cdot dc(A)$ on the left side and dt on the right side. The factors $-\frac{1}{v_A}$ and k are constants with respect to the integration and thus can be taken outside the integral. In order to calculate determined integrals, one also needs to specify the boundaries within which the integration is supposed to happen. We start at time t_0 where the concentration of A is the initial concentration $c_0(A)$, and proceed up to time t, where the concentration of A is $c_t(A)$

$$-\frac{1}{v_A}\int_{c0}^{c1} \frac{1}{c(A)} \cdot dc(A) = k \cdot \int_{t0}^{t} dt \tag{2.56}$$

The integral on the left side, $\int \frac{1}{c(A)} \cdot dc(A)$, corresponds to the type $\int x - 1dx$ and is resolved to ln$x$. On the right hand side, $\int dt$ is of the type $\int dx$ which resolves to x. Eq. (2.56) with determined integrals thus yields:

$$-\frac{1}{v_A}[\ln c(A)]_{c0}^{c_t} = k \cdot [t]_{t0}^{t} \tag{2.57}$$

Eq. (2.57) then yields:

$$-\frac{1}{v_A}[\ln c_t(A) - \ln c_0(A)] = k \cdot (t - t_0)$$

With $t_0 = 0$ one obtains:

$$-\frac{1}{v_A}[\ln c_t(A) - \ln c_0(A)] = k \cdot t$$

In order to calculate a numerical value for the only unknown, $c_t(A)$, we isolate the expression for $c_t(A)$ on one side:

$$\ln c_t(A) - \ln c_0(A) = - v_A \cdot k \cdot t \tag{2.58}$$

$$\ln c_t(A) = - v_A \cdot k \cdot t + \ln c_0(A) \tag{2.59}$$

Eq. (2.59) shows that if we measure the concentration of A, $c_t(A)$, at various time points t, and plot ln$c_t(A)$ versus t, we obtain a line with a negative slope, $(-v_A \cdot k)$, and a y-intercept of ln$c_0(A)$ (Fig. 2.1). Linear relationships in x-y plots are a convenient way of analysing data, and historically (before the advent of the computer) the only way to determine numerical values of fitting constants.

With modern fitting software, it is, of course, possible to fit non-linear relationships. If Eq. (2.58) is resolved for $c_t(A)$, one obtains:

$$\ln\frac{c_t(A)}{c_0(A)} = - v_A \cdot k \cdot t \tag{2.60}$$

$$\frac{c_t(A)}{c_0(A)} = e^{-v_A \cdot k \cdot t}$$

$$c_t(A) = c_0(A) \cdot e^{-v_A \cdot k \cdot t} \tag{2.61}$$

A closer look at Eqs. (2.59) and (2.61) reveals that the experimentally determined rate constant is indeed ($v_A \cdot k$) and includes the stoichiometric coefficient! This needs to be taken into account when calculating the rate constant k from experimental data.

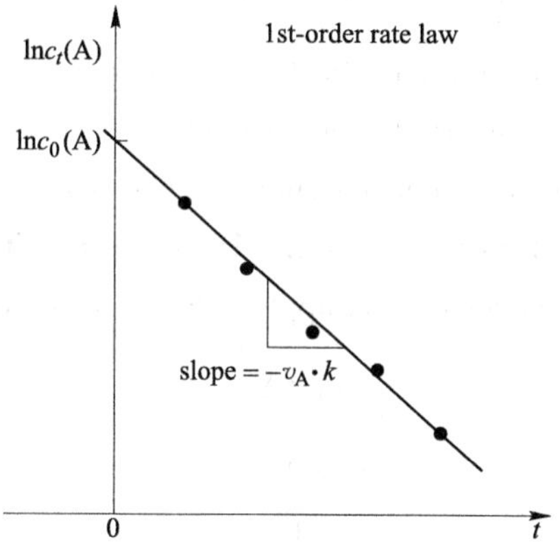

Fig. 2.1 Analysis of 1st-order kinetics by means of linear regression

2.2.2.3 Half-lives and Time Constants

Half-lives and ***time constants*** are secondary parameters derived from the rate equations and can be calculated for reactions of any order.

The half-life $t_{1/2}$ describes the time required to reduce the concentration of a reactant to 1/2 of its initial value. The time constant τ describes the time required to reduce the concentration of a reactant to 1/e of its initial value.

For 1st-order chemical reactions, the half-life and time constant are particularly useful indicators for the rate of the reaction, since with 1st-order reactions the two parameters do not depend on the initial concentration of reactants is the half-life $t_{1/2}$ of a reactant.

In order to derive an expression for half-life of a 1st-order reaction, we consider its definition:

$t = 0$: $$c_t(A) = c_0(A)$$

$t = t_{1/2}$: $$c_t(A) = \frac{1}{2} c_0(A)$$

Using this relationship in the 1st-order rate:

$$\ln \frac{c_t(A)}{c_0(A)} = -k \cdot t$$

$$\ln \frac{\frac{1}{2} c_0(A)}{c_0(A)} = -k \cdot t_{1/2}$$

$$-\ln \frac{1}{2} = k \cdot t_{1/2}$$

Making use of a logarithm rule that multiplies (−1) into the argument of the logarithm yields:

$$\ln 2 = k \cdot t_{1/2}$$

and thus

$$t_{1/2} = \frac{\ln 2}{k} \quad \text{Half-life of a 1st-order reaction} \tag{2.62}$$

Irrespective of the initial concentration, the time it takes in a 1st-order reaction to reduce the initial concentration to half its value is always $0.693/k$.

Earlier, we have introduced the time constant τ of a reaction as the time it takes for the concentration of a reactant to fall to 1/e (e ≈ 2.71828) of its initial value. As above in the case of the half-life, we use this relationship and apply it to the 1st-order rate law which yields:

$$\ln \frac{\frac{1}{e} c_0(\mathrm{A})}{c_0(\mathrm{A})} = -k \cdot \tau$$

$$-\ln \frac{1}{e} = k \cdot \tau$$

$$\ln e = k \cdot \tau$$

$$1 = k \cdot \tau$$

$$\tau = 1/k \quad \text{Time constant of a 1st-order reaction} \tag{2.63}$$

2.2.2.4 Integrated Rate Expression for the Second Order Rate Law

The differential 2nd-order rate law is given by Eq. (2.64):

$$-\frac{1}{v_\mathrm{A}} \cdot \frac{\mathrm{d}c(\mathrm{A})}{\mathrm{d}t} = k \cdot c(\mathrm{A})^2 \tag{2.64}$$

We re-arrange the equation such that the two variables $c(\mathrm{A})$ and t get isolated on opposite sides:

$$-\frac{1}{v_\mathrm{A}} \cdot \frac{1}{c(\mathrm{A})^2} \cdot \mathrm{d}c(\mathrm{A}) = k \cdot \mathrm{d}t$$

and integrate from time t_0 where the concentration of A is $c_0(\mathrm{A})$ to time t, where the concentration of A is $c_t(\mathrm{A})$:

$$-\frac{1}{v_\mathrm{A}} \cdot \int_{c_0}^{c_t} \frac{\mathrm{d}c(\mathrm{A})}{c(\mathrm{A})^2} = k \cdot \int_{t_0}^{t} \mathrm{d}t \tag{2.65}$$

The integral on the left side, $\int \frac{\mathrm{d}c(\mathrm{A})}{c(\mathrm{A})^2}$ corresponds to the type $\int x^{-2}\mathrm{d}x$ and is resolved to $\frac{1}{-2+1} \cdot x^{-2+1} = -x^{-1}$. On the right hand side, $\int \mathrm{d}t$ is of the type $\int \mathrm{d}x$ which resolves to x. Eq. (2.65) with determined integrals thus yields:

$$\frac{1}{v_\mathrm{A}} \cdot \left[\frac{1}{c(\mathrm{A})}\right]_{c_0}^{c_t} = k \cdot [t]_{t_0}^{t}$$

$$\frac{1}{v_A} \cdot \left[\frac{1}{c_t(A)} - \frac{1}{c_0(A)}\right] = k \cdot (t - t_0)$$

With $t_0 = 0$, this yields:

$$\frac{1}{c_t(A)} - \frac{1}{c_0(A)} = v_A \cdot k \cdot t$$

$$\frac{1}{c_t(A)} = v_A \cdot k \cdot t + \frac{1}{c_0(A)} \quad (2.66)$$

Eq. (2.66) shows that for the second order rate law, a linear correlation is obtained when plotting $1/c_t(A)$ versus t, where the slope is $v_A \cdot k$ and the y-intercept is $1/c_0(A)$ (Fig. 2.2).

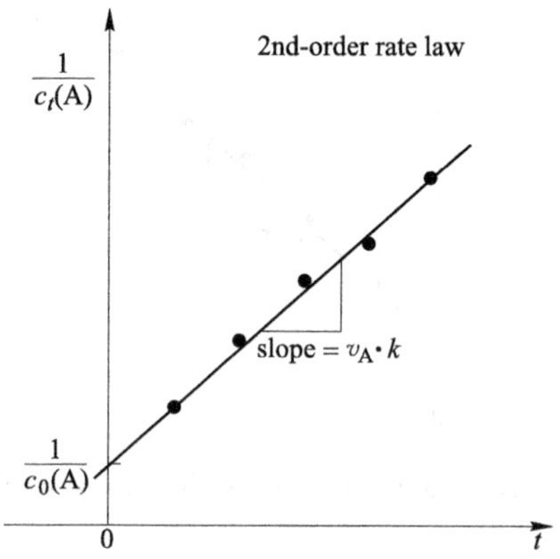

Fig. 2.2 Analysis of 2nd-order kinetics by means of linear regression

2.2.2.5 Integrated Rate Expression for the Third Order Rate Law

The differential 3rd-order rate law is given by Eq. (2.67):

$$-\frac{1}{v_A} \cdot \frac{dc(A)}{dt} = k \cdot c(A)^3 \quad (2.67)$$

For integration, we again isolate the two variables, $c(A)$ and t, on opposite sides:

$$-\frac{1}{v_A} \cdot \frac{1}{c(A)^3} \cdot dc(A) = k \cdot dt$$

and integrate from $[t_0, c_0(A)]$ to $[t, c_t(A)]$:

$$-\frac{1}{v_A} \cdot \int_{c_0}^{c_t} \frac{dc(A)}{3c(A)^3} = k \cdot \int_{t_0}^{t} dt \quad (2.68)$$

The integral on the left side, $\int_{c_0}^{c_t} \frac{dc(A)}{c(A)^3}$ corresponds to the type $\int x^{-3}dx$ and is resolved to $\frac{1}{-3+1} \cdot x^{-3+1} = -\frac{1}{2}x^{-2}$. On the right hand side, $\int_{t_0}^{t} dt$ is of the type $\int dx \int dx$ which resolves to x.

Eq. (2.68) with determined integrals thus yields:

$$-\frac{1}{v_A} \cdot \left[-\frac{1}{2 \cdot c(A)^2}\right]_{c_0}^{c_t} = k \cdot [t]_{t_0}^{t}$$

$$\frac{1}{2 \cdot v_A} \cdot \left[\frac{1}{c_t(A)^2} - \frac{1}{c_0(A)^2}\right] = k \cdot (t - t_0)$$

$$\frac{1}{c_t(A)^2} = 2 \cdot v_A \cdot k \cdot t + \frac{1}{c_0(A)^2} \tag{2.69}$$

Therefore, a plot of $\frac{1}{c_t(A)^2}$ against t will yield a line with slope $2 \cdot v_A \cdot k$ and intercept $\frac{1}{c_0(A)^2}$ (Fig. 2. 3).

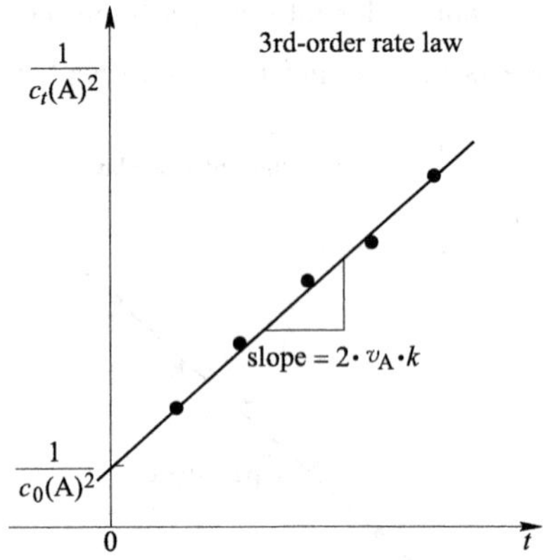

Fig. 2. 3 Analysis of 3rd-order kinetics by means of linear regression

2. 2. 2. 6 Integrated Rate Expression for the Zero-th Order Rate Law

In order to integrate ***the differential rate law of zero-th order***

$$-\frac{1}{v_A} \cdot \frac{dc(A)}{dt} = k$$

the two variables, $c(A)$ and t, are isolated on opposite sides of the equation:

$$-\frac{1}{v_A} \cdot dc(A) = k \cdot dt$$

We then integrate from $[t_0, c_0(A)]$ to $[t, c_t(A)]$:

$$-\frac{1}{v_A} \cdot \int_{c_0}^{c_t} dc(A) = k \cdot \int_{t_0}^{t} dt \tag{2.70}$$

The integrals on both sides side, $\int_{c_0}^{c_t} dc(A)$ and $\int_{t_0}^{t} dt$, correspond to the type $\int dx$ which resolves to x. Eq. (2. 70) with determined integrals thus yields:

$$-\frac{1}{v_A} \cdot [c(A)]_{c_0}^{c_t} = k \cdot [t]_{t_0}^{t}$$

which results in

$$-\frac{1}{v_A} \cdot [c_t(A) - c_0(A)] = k \cdot (t - t_0)$$

With $t_0 = 0$ this yields:

$$-\frac{1}{v_A} \cdot [c_t(A) - c_0(A)] = k \cdot t$$

$$c_t(A) - c_0(A) = - v_A \cdot k \cdot t$$

$$c_t(A) = - v_A \cdot k \cdot t + c_0(A) \quad (2.71)$$

Therefore, a plot of $c_t(A)$ against t will yield a line with slope $-v_A \cdot k$ and intercept $c_0(A)$ (Fig. 2.4).

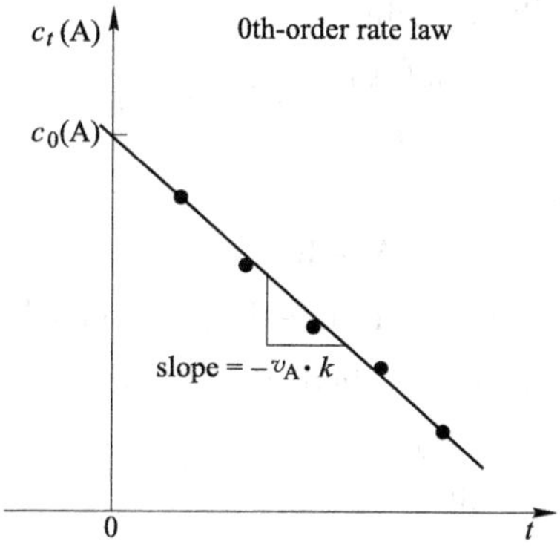

Fig. 2.4 Analysis of 0th-order kinetics by means of linear regression

2.2.3 Linking the Rate Laws with Reaction Mechanisms

So far, we have focused on the explanation of kinetic observations with respect to mathematically formulated rate laws. In the following, we will try to explain the observed rate laws in terms of a postulated reaction mechanism.

2.2.3.1 Elementary Reactions

Most reactions occur as a sequence of steps which are called elementary reactions, each of which involves only a small number of different molecules or ions.

For example, the net reaction describing the formation of HBr from its elements

$$H_2(g) + Br_2(g) \longrightarrow 2HBr(g)$$

can be broken down into the elementary reactions arising from a radical reaction mechanism:

Initiation	$Br_2 \longrightarrow Br^{\cdot} + Br^{\cdot}$
Propagation	$Br^{\cdot} + H_2 \longrightarrow HBr + H^{\cdot}$
	$H^{\cdot} + Br_2 \longrightarrow HBr + Br^{\cdot}$
	$H^{\cdot} + HBr \longrightarrow Br^{\cdot} + H_2$
Termination	$Br^{\cdot} + Br^{\cdot} \longrightarrow Br_2$
	$H^{\cdot} + Br^{\cdot} \longrightarrow HBr$

In an elementary reaction, there is no phase specified; it is simply a proposed individual step of a larger reaction mechanism. For instance in the following step

$$H^{\cdot} + Br_2 \longrightarrow HBr + Br^{\cdot}$$

an H atom attacks a Br_2 molecule: a bimolecular reaction. It was also discussed that the molecularity needs to be distinguished from the reaction order which is an empirical quantity, derived from the experimental rate law. However, the reaction order of the individual elementary reactions can be derived directly from the molecularity.

Unimolecular Elementary Reactions

$$A \longrightarrow P$$

follow a 1st-order rate law, because the number of molecules A that can decay is proportional to the number of molecules A initially available:

$$\frac{dc(A)}{dt} = -k \cdot c(A) \tag{2.72}$$

Bimolecular Elementary Reactions

$$A + B \longrightarrow P$$

follow a 2nd-order rate law, because their rate is proportional to the rate with which the two reactants meet—which in turn depends on their concentrations:

$$\frac{dc(A)}{dt} = -k \cdot c(A) \cdot c(B) \tag{2.73}$$

2.2.3.2 Consecutive Elementary Reactions

When investigating a reaction mechanism, the postulated mechanism can only be explored by a detailed investigation of the system, considering that side products or intermediates may appear in the course of the reaction. Importantly, if a reaction is an elementary bimolecular reaction, then it has a 2nd-order kinetics. However, if 2nd-order kinetics are observed for a net reaction, then it may be a complex reaction.

In the following, we will need to combine a series of consecutive simple steps, i. e. elementary reactions, in order to arrive at a reaction mechanism and the corresponding rate law.

Some reactions proceed through formation of an intermediate (I), so the entire mechanism is described by consecutive unimolecular reactions:

$$A \xrightarrow{k_1} I \xrightarrow{k_2} P$$

An example is the decay of radioactive isotopes, such as

$$^{239}U \xrightarrow{t\frac{1}{2} = 23.5\text{min}} {}^{239}Np \xrightarrow{t\frac{1}{2} = 2.35\text{d}} {}^{239}Pu$$

When we consider the variation of concentration with time for consecutive unimolecular reactions, it becomes clear that A decays according to a 1st-order rate law to I:

$$\frac{dc(A)}{dt} = -k_1 \cdot c(A)$$

The intermediate I is formed from A, and decays to P; both processes follow 1st-order rate laws:

$$\frac{dc(I)}{dt} = k_1 \cdot c(A) - k_2 \cdot c(I)$$

P is formed from I in a 1st-order reaction

$$\frac{dc(P)}{dt} = k_1 \cdot c(I)$$

In order to obtain expressions for the concentrations of A, I and P, the differential equations need to be integrated. This yields for the above equations:

$$c(A) = c_0(A) \cdot e^{k_1 \cdot t} \tag{2.74}$$

$$c(I) = \frac{k_1}{k_2 - k_1} \cdot (e^{-k_1 \cdot t} - e^{-k_2 \cdot t}) \cdot c_0(A) \tag{2.75}$$

$$c(P) = \left(1 + \frac{k_1 \cdot e^{-k_2 \cdot t} - k_2 \cdot e^{-k_1 \cdot t}}{k_2 - k_1}\right) \cdot c_0(A) \tag{2.76}$$

The time-dependent development of the concentrations of A, I and P is illustrated graphically in Fig. 2.5.

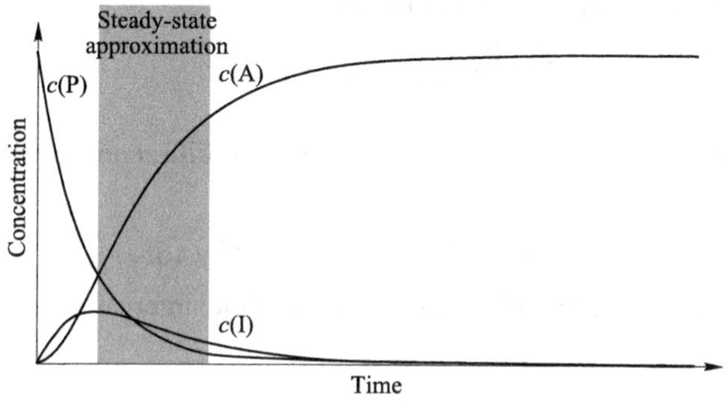

Fig. 2.5 Concentrations of reactant, intermediate and product in a set of consecutive elementary interactions

$$c(I) \approx \text{const.} \rightarrow \frac{dc(I)}{dt} \approx 0 \tag{2.77}$$

2.2.3.3 Steady-State Approximation

It is quite apparent from the algebraic expressions (2.74) ~ (2.76) in the previous section, that there is an increase in mathematical complexity when looking at reactions with more than one step. A reaction scheme that comprises many steps may become virtually unsolvable, and numerical rather than algebraic integration may be required.

Alternatively, an approximation can be introduced that yields more directly accessible results than the rigorous mathematical treatment. ***The (quasi-) steady-state*** approximation assumes that, after an initial induction period, where concentrations of intermediates rise from zero to a peak value, they remain approximately constant and the rate of change of concentrations of intermediates are negligibly small.

Fig. 2.5 shows that this is indeed an approximation as the curve depicting the concentration of the intermediate I does not possess a slope $\left(\frac{dc(I)}{dt}\right)$ of zero after rising to a peak value. Nevertheless, we assume the approximation stated by Eq. (2.76) for the following considerations and obtain for the concentration of the intermediate:

$$\frac{dc(I)}{dt} = k_1 \cdot c(A) - k_2 \cdot c(I) = 0$$

and thus

$$c(I) = \frac{k_1}{k_2} \cdot c(A)$$

which yields for the rate of product concentration change:

$$\frac{dc(P)}{dt} = k_2 \cdot c(I) = \frac{k_1 \cdot k_2}{k_2} \cdot c(A) = k_1 \cdot c(A)$$

This reveals a 1st-order rate law, whereby P is formed by a decay process of A:

$$c(A) = c_0(A) \cdot e^{-k_1 \cdot t} \tag{2.78}$$

For the rate of product formation, one thus obtains:

$$\frac{dc(P)}{dt} = k_1 \cdot c_0(A) \cdot e^{-k_1 \cdot t} \tag{2.79}$$

In order to integrate Eq. (2.79), we first isolate the two differential variables on opposite sides of the equation:

$$dc(P) = k_1 \cdot c_0(A) \cdot e^{-k_1 \cdot t} \cdot tdt$$

and then consider the development of products in the time interval $0 \rightarrow t$, i. e. from $c_0(P)$ to $c_t(P)$:

$$\int_{c_0(P)}^{c_t(P)} dc(P) = \int_0^t k_1 \cdot c_0(A) \cdot e^{-k_1 \cdot t} dt$$

The factors k_1 and $c_0(A)$ are constants with respect to the integral and can thus be taken outside the integral:

$$\int_{c_0(P)}^{c_t(P)} dc(P) = k_1 \cdot c_0(A) \cdot \int_0^t e^{-k_1 \cdot t} dt$$

The integral on the left side of the equation is of the type $\int dx = \int x^0 \, dx$ and resolves to $x^{0+1} = x$.

On the right hand side, the integral is of the type $\int e^{a \cdot x} dx$ which resolves to $1/a \cdot e^{a \cdot x}$. We thus obtain:

$$[c(P)]_{c_0(P)}^{c_t(P)} = k_1 \cdot c_0(A) \cdot \left[-\frac{1}{k_1} e^{-k_1 \cdot t} \right]_0^t$$

$$c_t(P) - c_0(P) = -\frac{k_1}{k_1} \cdot c_0(A) \cdot (e^{-k_1 \cdot t} - e^{-k_1 \cdot 0})$$

Since there is no product present at time $t=0$, it follows that $c_0(P) = 0$; for the term $e^{-k_1 \cdot 0}$, one obtains $e^0 = 1$. Therefore:

$$c_t(P) - 0 = -c_0(A) \cdot (e^{-k_1 \cdot t} - 1)$$

$$c_t(P) = c_0(A) \cdot (1 - e^{-k_1 \cdot t}) \tag{2.80}$$

When compared to the expression we obtained for the product concentration by rigorous kinetic analysis above

$$c(P) = \left(1 + \frac{k_1 e^{-k_2 \cdot t} - k_2 \cdot e^{-k_1 \cdot t}}{k_2 - k_1}\right) \cdot c_0(A) \tag{2.81}$$

it becomes clear that expression 2.80—obtained with the steady-state approximation—constitutes a special case of Eq. (2.81). In the case when

$$k_2 \gg k_1$$

k_1 in the denominator of the quotient in (2.81) can be neglected. Also, a large value of k_2 will give rise to $e^{-k_2 \cdot t}$ approaching zero, based on the behaviour of the function e^{-x} (see Fig. 2.6). Under these conditions, Eq. (2.81) becomes:

$$c(P) = \left(1 + \frac{k_1 \cdot 0 - k_2 \cdot e^{-k_1 \cdot t}}{k_2}\right) \cdot c_0(A)$$

$$c(P) = \left(1 + \frac{-k_2 \cdot e^{-k_1 \cdot t}}{k_2}\right) \cdot c_0(A)$$

$$c(P) = (1 - e^{-k_1 \cdot t}) \cdot c_0(A)$$

and thus shows the relationship (2.80) obtained with the steady-state approximation. It is found that k_2 does not have to be much larger than k_1 in order to obtain reasonably accurate results from the steady-state approximation. For example, $k_2 = 20 \cdot k_1$ already yields a good agreement.

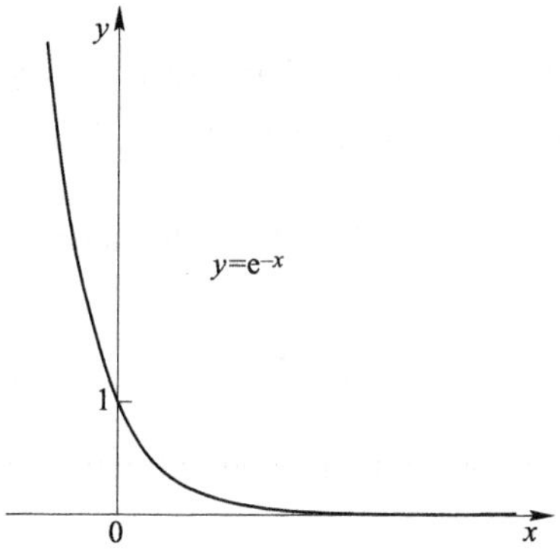

Fig. 2.6 The function e^{-x} anneals to zero when x assumes very large values

2.2.3.4 The Rate-Determining Step

In the previous section, we have considered the special case of $k_2 \gg k_1$ in the following set of consecutive reactions

$$A \xrightarrow{k_1} I \xrightarrow{k_2} P$$

which means that the first reaction proceeds a lot slower than the second reaction. The first reaction thus becomes ***the rate-determining step***.

In consecutive reactions, the rate-determining step is the one with the lowest rate constant. In many cases, the rate law for a consecutive reaction with a ratedetermining step can thus be written in a straight-forward way: the rate of the overall reaction equals the rate of the rate-determining step.

2.2.3.5 Pre-Equilibria

In many instances of consecutive reactions, the first reaction needs to be treated as an equilibrium process (pre-equilibrium). We thus consider the case where an intermediate I is formed by reaction of A and B in an equilibrium reaction, and that the reaction to the product P proceeds a lot slower than the reverse reaction in the equilibrium:

$$A + B \xleftrightarrow{k_{\pm 1}} I \xrightarrow{k_2} P$$

Since we assume the product formation as rate-determining step, the rate constant for the second reaction takes a much smaller value than the rate constant for the reverse reaction of the equilibrium:

$$k_{-1} \gg k_2$$

For the first reaction, the equilibrium constant is given by:

$$K = \frac{[I]}{[A] \cdot [B]} = \frac{c(I) \cdot c^{\ominus}}{c(A) \cdot c(B)} = \frac{k_1}{k_{-1}} \cdot c^{\ominus}$$

Substituting this into the rate law for the product formation yields:

$$\frac{dc(P)}{dt} = k_2 \cdot c(I) = k_2 \cdot \frac{K}{c^{\ominus}} \cdot c(A) \cdot (B) \qquad (2.82)$$

This is the form of a 2nd-order rate law with the rate constant $k = k_2 \cdot \frac{K}{c^{\ominus}} = k_2 \cdot \frac{k_1}{k_{-1}}$.

2.2.3.6 Kinetic and Thermodynamic Control

A common feature of reactions in organic chemistry (e.g. nitration of monosubstitute benzenes yielding *o*-, *m*-, *p*-derivatives) is the formation of multiple products from one set of reactants:

$$A+B \rightarrow P_1 \quad \text{rate constant: } k_1; \text{ rate: } v_1 = \frac{dc(P_1)}{dt} = k_1 \cdot c(A) \cdot c(B)$$

$$A+B \rightarrow P_2 \quad \text{rate constant: } k_2; \text{ rate: } v_2 = \frac{dc(P_2)}{dt} = k_2 \cdot c(A) \cdot c(B)$$

The relative proportion of products formed is given by the ratio of the two rates, and thus the two rate constants:

$$\frac{v_1}{v_2} = \frac{k_1}{k_2} = \frac{c(P_1)}{c(P_2)} \qquad (2.83)$$

The ration given by Eq. (2.83) is due to the kinetic control over the proportion of products formed. Notably, this refers to the concentration of products produced at any stage during the reaction, i.e. before the reaction reached equilibrium. Kinetic control is also the reason for the occurrence of electrochemical overpotentials; If a reaction is allowed to reach equilibrium, then the proportion of products is determined by the thermodynamic considerations. The ratio is then under thermodynamic control.

2.3 Electrochemical Foundation

2.3.1 Electrolyte Solution

2.3.1.1 Ions in Solution

In the previous sections, we have mostly assumed that the systems under study consisted of non-dissociating solutes, i.e. non-electrolytes. We now want to expand considerations to substances that dissociate in solution. Such substances are called electrolytes and form ions when being dissolved in solvents. Electrolytes can be classified into strong and weak electrolytes.

Strong electrolytes completely dissociate in solution:

$$NaCl(s) + H_2O \longrightarrow Na^+(aq) + Cl^-(aq) + H_2O$$

$$HCl(g) + H_2O \longrightarrow H_3O^+(aq) + Cl^-(aq)$$

$$AgCl(s) + H_2O \longrightarrow Ag^+(aq) + Cl^-(aq) + H_2O$$

In contrast, weak electrolytes do not fully dissociate in solution, due to inter-ionic interactions:

$$H_3C-COOH(l) + H_2O \longrightarrow H_3O^+(aq) + H_3C-COO^-(aq)$$

The occurrence of ions in solutions of electrolytes is not dependent on the flow of current; electrolytes dissociate readily upon dilution in solvent. The degree of dissociation α describes the fraction of solute present as ions.

In aqueous solution, ions have water molecules associated with them; this is called the hydration shell. Ions change the structure of the water hydrogen bond network. In the presence of an ion, a water molecule reorients such that its polarized charge faces the opposite charge of the ion. During this reorientation process, the hydrogen bonds of the water molecule to its nearest neighbor is broken. The orientation of the water molecules that form the hydration shell directly around the ion is called the inner solvation shell (see Fig. 2.7, orange molecules) and results in a net charge on the outside of this shell. This charge is of the same sign as that of the ion in the center. The charge on the outside of this inner hydration shell causes further water molecules in the vicinity to reorient, leading to a second solvation shell (see Fig. 2.7, red molecules)

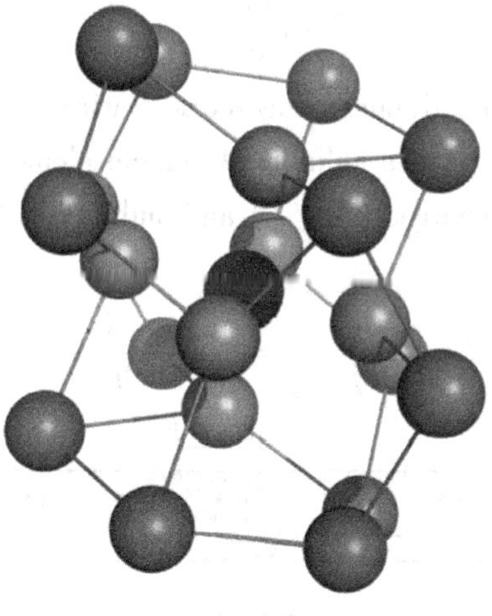

Scan the QR code to see color picture

Fig. 2.7 Quantum-chemical calculations with the semi-empirical AM1 method show that monoatomic ions (such as e.g. Na^+, blue) exist in a hydration shell resulting in the complex $[Na(OH_2)_{20}]^+$ (Peslherbe et al., 2000)

(The shell consists of an inner solvation shell where six water molecules take the vertex positions of an octahedron (orange); The remaining 14 water molecules (red) form the second solvation shell; The water structure around the cation takes the form of a puckered dodecahedron; The formation of this structure can be thought of as a 'pulling in' of the inner solvation shell molecules from their initial positions on the vertices of a regular dodecahedron)

The occurrence of such hydration shells explains the freezing point depression of solutions. The hydration shells of dissolved ions disrupt the hydrogen bonding network of water that would otherwise form the hexagonal structure of ice.

2.3.1.2 Charge and Electroneutrality

The charge Q describes ***the quantity of electricity***; it can be positive (cations, protons) or negative (anions, electrons). The charge is measured in units of Coulomb:

$$[Q] = 1\text{C}$$

If matter possesses only fixed charges, then it is called an ***insulator***. In contrast, the existence of mobile charges makes matter a ***conductor***. In case of an electronic conductor, the mobile charges are electrons, in ionic conductors, the mobile charges are ions.

The principle of electroneutrality states that there can be no significant net charge in any macroscopic volume within a conductor. This is a consequence of the work required to separate opposite charges, or to bring like charges into close contact. This work raises the free energy change of the underlying process that may lead to unbalanced charges, thus making it less spontaneous. The amount of unbalanced charges that is allowed is due to different concentrations of oppositely charged species that are not chemically significant.

2.3.1.3 Electric Current

Current (***I***) is the flow of charge dQ in a particular time interval dt through a defined volume:

$$I = \left(\frac{\mathrm{d}Q}{\mathrm{d}t}\right)_{\text{volume}} \tag{2.84}$$

If we consider charges moving through a cylindrical wire (Fig. 2.8), we can calculate the amount of charge dQ passing through a cylindrical volume element during the time interval dt by counting the number of passing charges (N_{passing}) and multiply with the unit charge e as well as the charge state z:

$$I = \left(\frac{\mathrm{d}Q}{\mathrm{d}t}\right)_{\text{volume}} = \frac{N_{\text{passing}} \cdot (z \cdot e)}{\mathrm{d}t} \tag{2.85}$$

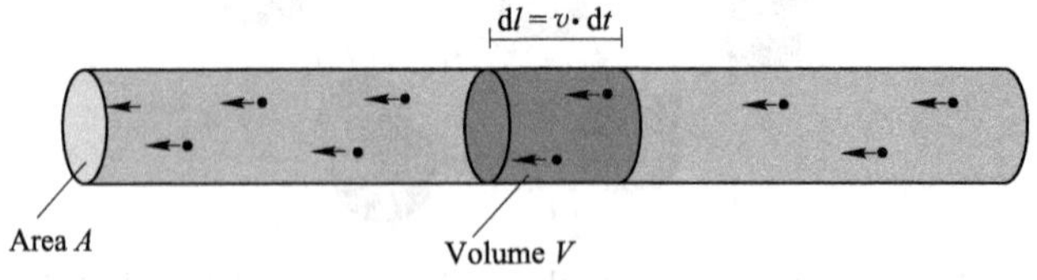

Fig. 2.8 Mobile charges passing through a cylindrical volume element V

By multiplying and diving with the volume V of the volume element and substituting

$$N_{\text{passing}} = n \cdot N_{\text{A}} \quad \text{and} \quad V = A \cdot \mathrm{d}l$$

we obtain:

$$I = \frac{\dfrac{N_{\text{passing}}}{V} \cdot V(z \cdot e)}{\mathrm{d}t} = \frac{\dfrac{n \cdot N_{\text{A}}}{V} \cdot A \cdot \mathrm{d}l \cdot (z \cdot e)}{\mathrm{d}t}$$

We can further substitute

$$c = n/V \quad \text{and} \quad dl = v \cdot dt$$

which yields

$$I = \frac{\frac{N_{\text{passing}}}{V} \cdot V(z \cdot e)}{dt} = \frac{c \cdot N_A \cdot A \cdot v \cdot dt \cdot (z \cdot e)}{dt}$$

This simplifies to:

$$I = \frac{c \cdot N_A \cdot A \cdot v \cdot dt \cdot (z \cdot e)}{dt} = z \cdot c \cdot N_A \cdot A \cdot v \cdot e = z \cdot c \cdot A \cdot v = z \cdot c \cdot A \cdot v \cdot F \tag{2.86}$$

F is the Faraday constant and has the value of:

$$F = N_A \cdot e = 6.022 \times 10^{23}\ \text{mol}^{-1} \times 1.602 \times 10^{-19}\text{C} = 96485\text{C/mol}$$

Eq. (2.86) provides a relation between the current and the molar concentration c of charged particles with the charge state z that move with the speed v.

2.3.1.4 Electric Potential

Fig. 2.9 shows the scheme of a simple electric circuit where a direct current power supply gives rise to electrons moving from the cathode (excess of electrons) to the anode (shortage of electrons) through a cylindrical wire. The fact that electrons move from the cathode to the anode can be conceptualized by the existence of a potential difference ΔE between the cathode and the anode. The potential difference ΔE is measured in units of volts:

$$[\Delta E] = 1\text{V}$$

Whereas the physical movement of electrons has the direction cathode→anode, the current I is defined to flow in the opposite direction.

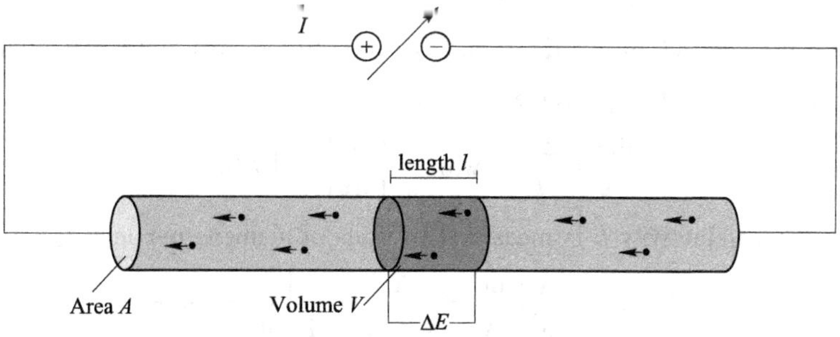

Fig. 2.9 Illustration to derive the relationship between the potential difference ΔE, current I and the travelling parameters (length, cross-sectional area) of the moving charges

As in the previous section, we consider a volume element V in the cylindrical wire of Fig. 2.9; the volume of this element can be calculated from the cross-sectional area A and the length l, as per $V=A \cdot l$.

It is obvious that there will be more current flowing, if the potential difference of the power supply is higher; I and ΔE will thus be proportional to each other. Because the potential difference between the anode and the cathode is taken as positive, and the current flows from the anode to the cathode, the directions of the two phenomena are opposite, hence:

$$\Delta E \sim -I \tag{2.87}$$

If we increase the length l of the volume element, we will need to increase the potential difference ΔE to have the same amount of charges (i. e. the same current) flowing through that volume element. Therefore:

$$\Delta E \sim l \tag{2.88}$$

The opposite is true for the cross-sectional area A. If we widen the area A, but require the same amount of charges to flow through, the potential difference ΔE needs to be decreased. It thus appears that

$$\Delta E \sim \frac{1}{A} \tag{2.89}$$

It follows from Eqs. (2.87) ~ (2.89):

$$\Delta E \sim \frac{-I \cdot l}{A} \tag{2.90}$$

If we define a new quantity j called **the current density** as

$$j = \frac{I}{A} \tag{2.91}$$

then we obtain from Eq. (2.91):

$$j = \frac{I}{A} \sim \frac{\Delta E}{l} \tag{2.92}$$

As proportionality factor we introduce k as the **electric conductivity**

$$j = \frac{I}{A} = -k \cdot \frac{\mathrm{d}\phi}{\mathrm{d}x} = 1 \quad \text{with} \quad [j] = 1\mathrm{A/m} \tag{2.93}$$

The potential difference ΔE over a distance l is called the electric field. The electric field is the gradient of the electric potential ϕ over a distance:

$$\frac{\mathrm{d}\phi}{\mathrm{d}x} = \frac{\Delta E}{l} \quad \text{with} \quad \left[\frac{\mathrm{d}\phi}{\mathrm{d}x}\right] = 1\mathrm{V/m} \tag{2.94}$$

It follows that the conductivity k is measured in units of siemens per metre:

$$[k] = 1 \cdot \frac{\mathrm{A \cdot m}}{\mathrm{m^2 \cdot V}} = \frac{\mathrm{A}}{\mathrm{V \cdot m}} = \frac{1}{\Omega \cdot \mathrm{m}} = 1\mathrm{S/m} \tag{2.95}$$

Since conductivity is a specific property of a substance, it lends itself as a quantity to distinguish conducting from insulating matter.

2.3.1.5 Resistance

An important relationship for electric circuits (see Fig. 2.9) is that between the electric potential and the current flowing through the circuit. From Eq. (2.96) we can resolve an expression for the potential ΔE

$$j = \frac{I}{A} = -k \cdot \frac{\Delta E}{l} \tag{2.96}$$

$$\Delta E = \frac{I \cdot l}{k \cdot A} \tag{2.97}$$

The quotient $\frac{l}{k \cdot A}$ with the length l, the area A and the conductivity k describes constants of a particular electric circuit and as such represents a material constant for the given system. This quotient is thus defined as ***the electric resistance R***:

$$R = \frac{l}{k \cdot A} \tag{2.98}$$

and Eq. (2.97) yields:

$$|\Delta E| = I \cdot R \tag{2.99}$$

which constitutes Ohm's law. The resistance is measured in units of ohm: $[R] = 1\Omega$.

The electric resistance of a conductor represents the opposition to the passage of charges through that conductor. Intuitively, the willingness of a conductor to let charges pass through will be a quantity that is inversely related to the resistance. The quantity describing the ease with which charges can pass through is called ***the conductance G***:

$$G = 1/R = I/\Delta E \tag{2.100}$$

The conductance is measured in units of siemens: $[G] = 1\Omega^{-1} = 1\text{S}$. Like the resistance, the conductance is a property of the conductor used in the electric circuit.

2.3.1.6 Conductivity and Conductance

The electric conductivity k is the ratio between the current density and the electric field [Eq. (2.93)]. As such, it is a property of the conducting material. In contrast, ***the conductance G*** describes the current carrying capacity of the electrolytic substance; in solutions of electrolytes, the conductance is thus a property of the dissolved ions. The conductance of electrolytic solutions increases with dilution for both strong and weak electrolytes, because at low concentration there is less hindrance for the migrating ions from neighbours.

The specific conductance g measures the current-carrying capacity of all ions in a specific volume:

$$g = G/V \text{, with} \quad [g] = 1\text{S/m}^3 \tag{2.101}$$

For strong electrolytes, the specific conductance decreases with dilution.

Whereas the conductance G depends on the physical size of the conductor, the conductivity k is independent of the conductor size. Conductivity and conductance are related as per:

$$j = \frac{I}{A} = -k \cdot \frac{\Delta E}{l} \tag{2.102}$$

$$\frac{|I| \cdot l}{A \cdot \Delta E} = k$$

With $G = 1/R = I/\Delta E$, it follows:

$$\frac{G \cdot l}{A} = k$$

$$G = K \cdot \frac{A}{l} \tag{2.103}$$

The conductance of a system is therefore equal to the conductivity of this system, multiplied by the area through which the migrating charges pass, and divided by the distance travelled.

2.3.1.7 Solubility Product

In the previous section, we have introduced the silver/silver chloride electrode as a frequently used reference electrode. The electrode is made of a silver wire coated with silver chloride that is immersed into a solution containing chloride ions. The step which determines the potential of this electrode (half-cell) is the transition of Ag^+ ions from the electrode into solution. Here, the metal ions ($Ag^+(s)$) are in a heterogeneous equilibrium between the solid electrode material ($AgCl(s)$) and the dissolved ions ($Ag^+(aq)$, $Cl^-(aq)$). A saturated aqueous AgCl solution contains only 13μM Ag^+ and 13μM Cl^- ions; the equilibrium of the following disassociation reaction is thus on the left hand side

$$AgCl(s) + H_2O \longrightarrow Ag^+(aq) + Cl^-(aq)$$

and results in an equilibrium constant of small value:

$$K = \frac{\frac{c(Ag^+)}{c^\ominus}\frac{c(Cl^-)}{c^\ominus}}{\frac{c(AgCl)}{c^\ominus}} = \frac{[Ag^+] \cdot [Cl^-]}{[AgCl]} \tag{2.104}$$

Remember that H_2O is omitted from calculation of the equilibrium constant, since it is neither consumed nor produced. The concentration of any solid substances are considered to be constant in solvation reactions, hence $c(AgCl)$ is equal to const. The silver chloride concentration can thus be multiplied into the equilibrium constant, yielding a new constant, ***the solubility product***, which is solely a function of the concentration of dissolved ions:

$$K_{sp} = K \cdot [AgCl] = [Ag^+] \cdot [Cl^-] = \frac{c(Ag^+)}{c^\ominus}\frac{c(Cl^-)}{c^\ominus} \tag{2.105}$$

The electrode (half-cell) potentials are themselves potential differences (see Fig. 2.10). They arise from the transition of ions from the solid to the dissolved state. To stay consistent with the nomenclature used in previous sections, we will denote these differences as E and $E^\ominus$, respectively, to indicate that we are referring to an electrochemical half-cell. We can formulate an expression for the potentials arising from Ag^+ transitions in the silver/silver chloride electrode, by using Eq. (2.105).

$$E(Cl^- | AgCl | Ag^+) = E^\ominus(Ag^+ | Ag) + \frac{R \cdot T}{z \cdot F} \cdot \ln \frac{[Ag^+(aq)]}{Ag^+(s)}$$

$$\rightarrow E(Cl^- | AgCl | Ag^+) = E^\ominus(Ag^+ | Ag) + \frac{R \cdot T}{F} \cdot \ln[Ag^+(aq)] \tag{2.106}$$

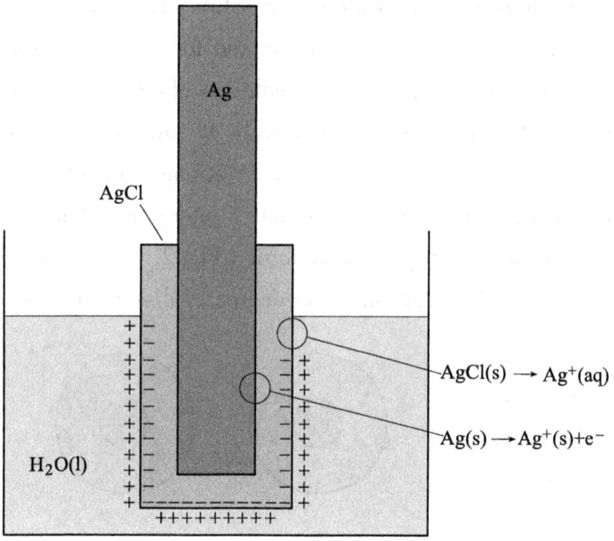

Fig. 2. 10 Illustration of transition processes at electrodes and the generation of an electrode potential

From Eq. (2. 105), we obtain

$$c_{aq}(Ag^+) = \frac{K_{sp}}{c_{aq}(Cl^-)} \cdot (c^{\ominus})^2$$

which we can substitute into Eq. (2. 106):

$$E(Cl^- \mid AgCl \mid Ag^+) = E^{\ominus}(Ag^+ \mid Ag) + \frac{R \cdot T}{F} \cdot \ln \frac{K_{sp}}{\frac{c_{aq}(Cl^-)}{c^{\ominus}}} \qquad (2.107)$$

Using the logarithm rule of $\frac{K_{sp}}{\frac{c_{aq}(Cl^-)}{c^{\ominus}}} = \ln K_{sp} - \ln \frac{c_{aq}(Cl^-)}{c^{\ominus}}$, we obtain:

$$E(Cl^- \mid AgCl \mid Ag^+) = E^{\ominus}(Ag^+ \mid Ag) + \frac{R \cdot T}{F} \cdot \ln K_{sp} - \frac{R \cdot T}{F} \cdot \ln \frac{c_{aq}(Cl^-)}{c^{\ominus}}$$

As $\frac{R \cdot T}{F} \cdot \ln K_{sp}$ is a constant, it can be combined with the standard potential of the $Ag^+ \mid Ag$ transition:

$$E(Cl^- \mid AgCl \mid Ag^+) = E^{\ominus}(Cl^- \mid AgCl \mid Ag^+) - \frac{R \cdot T}{F} \cdot \ln \frac{c_{aq}(Cl^-)}{c^{\ominus}} \qquad (2.108)$$

2. 3. 1. 8 Degree of Dissociation

If the deviation of the van't Hoff factor i from the theoretically expected value v is due to dissociation only, then the degree of ***dissociation*** $\boldsymbol{\alpha}$ can be determined as per:

$$\alpha = \frac{i - 1}{v - 1} \qquad (2.109)$$

where v is the theoretical number of ions yielded per solute molecule (e. g. for $CaCl_2$: $v=3$).

The degree of dissociation α can be reliably estimated for weak electrolytes, as there are only few ions at all concentrations, and therefore other solution effects are negligible.

For strong electrolytes, Eq. (2. 109) works well at low concentrations, but substantially deviates from experimental observations at moderate or strong concentrations, due to interactions between ions. With increasing concentrations, attractive interactions between dissolved electrolytes become important, and ion pairs or triplets may form (Fig. 2. 11). Since ions involved in those pairings no longer account as individual ions, the degree of dissociation is less than expected.

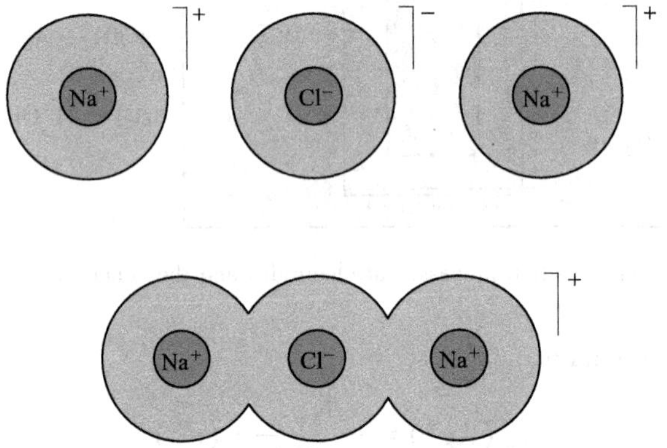

Fig. 2. 11 Illustration of ion association

(In solution, ions of opposite electrical charge may come together to form a distinct chemical entity. Top panel: fully solvated ions; bottom panel: an ion triplet with solvent sharing)

2. 3. 2 Electrolyte System

After having introduced some fundamental concepts and quantities of electricity, we will now consider their applications in a chemical context. Electrochemical reactions can proceed in systems when a suitable electric potential difference ΔE is applied. In this case, electric energy is converted into chemical energy. Such cells are called ***electrolytic cells***.

Alternatively, chemical reactions may result in the build-up of an electric potential difference, i. e. chemical energy is converted into electric energy. These cells are called ***galvanic cells***.

Conceptually, an electrochemical cell is separated into two half-cells, separating the oxidation and the reduction processes. The two electrochemical half reactions oxidation and reduction are combined into a net reaction, called redox reaction.

2. 3. 2. 1 Galvanic and Electrolytic Cells

If two half-cells are combined where the different chemical reactions give rise to an electric potential difference, the resulting electrochemical cell is called a galvanic cell. A prominent example is the so-called Daniell element shown in Fig. 2. 12, left panel. Here, the spontaneously proceeding redox reaction converts chemical into electric energy; this constitutes the working principle of batteries.

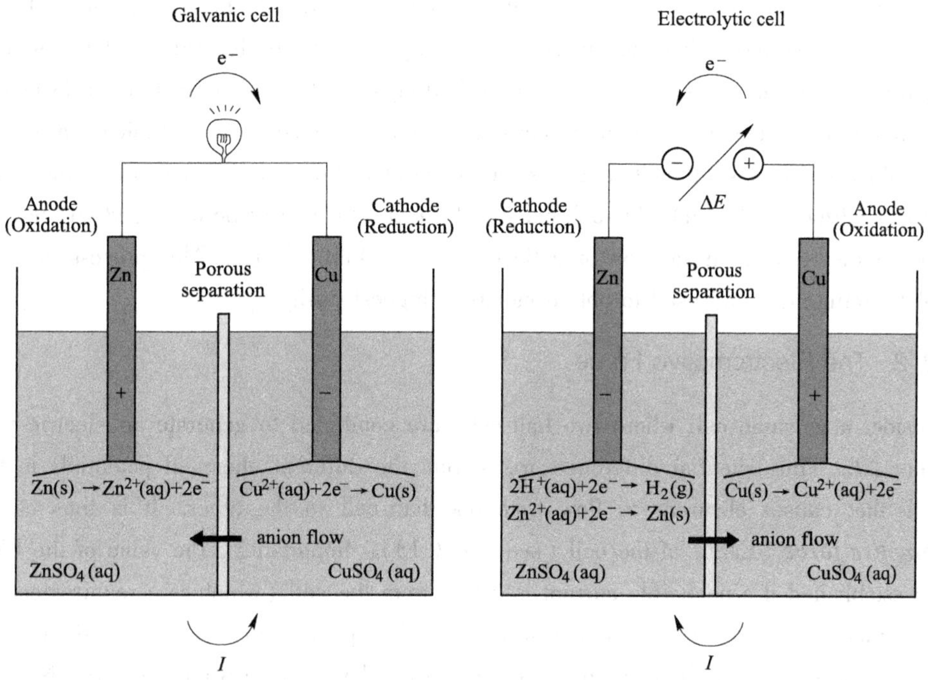

Fig. 2.12 Illustration of the processes in a galvanic cell (left panel) and electrolytic cell (right panel), using the Daniell element which consists of $Zn/ZnSO_4$ and $Cu/CuSO_4$ half-cells

According to Faraday, the definition of anode and cathode depend on charge and discharge. The anode is the electrode to which anions flow. The anode possesses positive potential and thus gives rise to an oxidation reaction, where suitable species (elementary zinc in Fig. 2.12, left panel) deposit their electrons onto the electrode. In contrast, the cathode is the electrode attracting cations; it is the electrode with negative potential. Here, suitable species (Cu^{2+} ions in Fig. 2.12, left panel) are reduced by receiving electrons from the electrode.

In an electrolytic cell, a chemical reaction is forced to occur due to current flowing through a cell. An electric potential needs to be applied externally (Fig. 2.12, right panel), causing conversion of electric to chemical energy. Both electrodes are placed in a container that contains the solution of molten or dissolved electrolyte. The external power supply provides the electrons. They enter the electrolyte solution through the cathode (negative potential) and leave through the anode (positive potential).

At the anode, which has a lack of electrons due to the positive potential, electrons are removed from suitable substances at the anode surface—an oxidation occurs. The cathode, in contrast, possesses a surplus of electrons due to its negative potential. Here, electrons will be transferred onto suitable substances—this is a reduction.

Fig. 2.12 contrasts the two different electrochemical cell types by using the same element, i.e. a $Zn/ZnSO_4$ and a $Cu/CuSO_4$ half-cell (the Daniell element). Of course, a galvanic cell is not supposed to be used as an electrolytic cell, hence the right panel of Fig. 2.12 is for illustration of the concept only. Note that when the Daniell element is inverted to become an electrolytic cell, the reduction at the anode does not deposit elementary zinc at the electrode, but rather produces hydrogen gas from water, since the reduction of hydrogen is electrochemically more favorable.

In technical applications, electrolytic cells are not constructed as two physical half-cells, but one cell which comprises the molten or dissolved electrolyte (e. g. $CuSO_4$) and electrodes made of the corresponding metal (Cu). In such cells, solid metal (Cu) will be deposited on the cathode; impurities either remain in solution or collect as an insoluble sludge. This process is known as electrolytic refinement and used to obtain metals of highest purity.

2. 3. 2. 2 The Electromotive Force

We consider a galvanic cell where two half-cells are combined to generate an electric potential difference ΔE. This potential difference arises from the different chemical potentials in the two half-cells that causes electrons to flow from one half-cell to the other. It is thus called ***the electromotive force*** (EMF) of the cell (see Fig. 2. 13). Importantly, the value of the EMF can only be established if a negligible current is drawn from the cell (which is a requirement for any Volt meter used to determine the voltage when placed in parallel to a resistor). If the two half-cells were connected by a short circuit (i. e. the external resistance would be zero, Rest = 0), then current flow would be maximized and there was no potential difference ($\Delta E = 0$). If the two half-cells are connected by a Volt meter which possesses a very large resistance, then the current I is negligible and the potential difference ΔE is maximized. This is the desired configuration, as any current flowing between the two half-cells is the consequence of a proceeding redox reaction and thus inevitably results in a lowering of the potential difference.

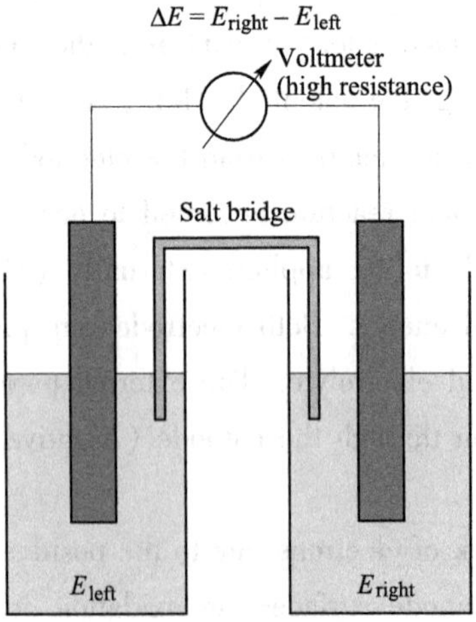

Fig. 2. 13 The convention for determining the electromotive force of an electrochemical cell

The combination of two half-cells is denoted by separating the electrode material and electrolyte solutions of the left and right half-cells by vertical lines. For example, the Daniell element, which we have introduced in Sect. 2. 3. 2. 1, is written as:

$$Zn(s) \mid 1M\ ZnSO_4 \mid 1M\ CuSO_4(aq) \mid Cu(s) \qquad EMF = \Delta E = +1.103V$$

The above cell denotes a zinc electrode immersed in an electrolyte solution of 1M aqueous $ZnSO_4$ as the left cell and a copper electrode immersed in 1M aqueous $CuSO_4$ solution as the right cell.

Importantly, the EMF(or electric potential difference) is by convention defined as

$$EMF = \Delta E = E_{right} - E_{left} \tag{2.110}$$

with E_{right} and E_{left} being the electric potentials of the right and left half-cells, respectively. In the example above, the right half-cell consisting of the system:

$$Cu^{2+}(aq) + 2e^- \longrightarrow Cu(s) \qquad E_{right} = 0.340V$$

possesses the higher potential and therefore pulls electrons from the left half-cell which consists of the system:

$$Zn^{2+}(aq) + 2e^- \longrightarrow Zn(s) \qquad E_{left} = 0.340V$$

Therefore, the spontaneously proceeding reactions in this electrochemical cell will consist of the reduction of Cu^{2+} ions and the oxidation of Zn, with the following net reaction:

$$Zn(s) + Cu^{2+}(aq) \longrightarrow Zn^{2+}(aq) + Cu(s)$$

$$\Delta E = E_{right} - E_{left} = 0.340V - (-0.763V) = 1.103V \tag{2.111}$$

The potential difference between the right and the left half-cell yields the electromotive force of that cell.

Importantly, when the number of electrons consumed or produced in the two half-cells differ, the redox potentials must not be multiplied. In manipulating potentials, one can only change the signs of the values, not the magnitude.

2.3.2.3 Combination of Two Half-Cells

In Sect. 2.3.2.2, we have introduced the combination of two half-cells as illustrated by the Daniell element:

$$Zn(s) \mid 1M\ ZnSO_4 \mid 1M\ CuSO_4(aq) \mid Cu(s) \qquad EMF = \Delta E = +1.103V$$

In this formalism, an electrochemical cell is denoted by a series of compartments. It is understood by convention that the electrode potential of the right-hand electrode is higher than that of the left-hand electrode, because the EMF is reported positive in above example, and calculated as

$$EMF = \Delta E = E_{right} - E_{left} \tag{2.112}$$

so if EMF> 0, then $E_{right} > E_{left}$.

If the cell above had been written in reverse order, then the EMF would be negative:

$$Cu(s) \mid 1M\ CuSO_4(aq) \mid 1M\ ZnSO_4(aq) \mid Zn(s) \qquad EMF = \Delta E = -1.103V$$

Importantly, the electrode with the higher potential is always the one where reduction occurs; the electrode with the lower potential is where oxidation occurs.

We can reformulate the above example in a more general fashion, combining a redox pair in the left half-cell (Red_{left}, Ox_{left}) with one in the right half-cell (Red_{right}, Ox_{right}):

$$Red_{left} \mid Ox_{left} \mid Ox_{right} \mid Red_{right} \qquad EMF = \Delta E > 0$$

with the following chemical reactions:

$$\text{Red}_{\text{left}} \longrightarrow \text{Ox}_{\text{left}} + ze^- \quad E_{\text{left}} \text{ oxidation}$$

$$\text{Ox}_{\text{right}} + ze^- \longrightarrow \text{Red}_{\text{right}} \quad E_{\text{right}} \text{ reduction}$$

$$\text{Red}_{\text{left}} + \text{Ox}_{\text{right}} \longrightarrow \text{Ox}_{\text{left}} + \text{Red}_{\text{right}} \quad \Delta E \text{ redox}$$

If $E_{\text{right}} > E_{\text{left}}$, then electrons will be flowing from the left half-cell to the right half-cell. That means ΔE is positive, and all reactions above will proceed from the left to right.

$$\Delta E = E_{\text{right}} - E_{\text{left}} \tag{2.113}$$

$$\Delta E = E^{\ominus}_{\text{right}} - \frac{RT}{zF} \cdot \ln \frac{c(\text{Red}_{\text{right}})}{c(\text{Ox}_{\text{right}})} - E^{\ominus}_{\text{left}} + \frac{RT}{zF} \cdot \ln \frac{c(\text{Red}_{\text{left}})}{c(\text{Ox}_{\text{left}})}$$

$$\Delta E = E^{\ominus}_{\text{right}} - E^{\ominus}_{\text{left}} + \frac{RT}{zF} \cdot \left[\ln \frac{c(\text{Red}_{\text{left}})}{c(\text{Ox}_{\text{left}})} - \ln \frac{c(\text{Red}_{\text{right}})}{c(\text{Ox}_{\text{right}})}\right]$$

$$\Delta E = \Delta E^{\ominus} + \frac{RT}{zF} \cdot \left[\ln \frac{c(\text{Red}_{\text{left}})}{c(\text{Ox}_{\text{left}})} + \ln \frac{c(\text{Ox}_{\text{right}})}{c(\text{Red}_{\text{right}})}\right]$$

$$\Delta E = \Delta E^{\ominus} + \frac{RT}{zF} \cdot \left[\ln \frac{c(\text{Red}_{\text{left}}) \cdot c(\text{Ox}_{\text{right}})}{c(\text{Ox}_{\text{left}}) \cdot c(\text{Red}_{\text{right}})}\right]$$

$$\Delta E = \Delta E^{\ominus} + \frac{RT}{zF} \cdot \ln \frac{1}{K}$$

$$\Delta E = \Delta E^{\ominus} - \frac{RT}{zF} \cdot \ln K \tag{2.114}$$

When the reactants of the cell have reached their equilibrium concentrations, there is no electric current flowing between the two half-cells (such as for example in a flat battery), and $\Delta E = 0$. It then follows that:

$$\Delta E = 0 = \Delta E^{\ominus} - \frac{RT}{zF} \cdot \ln K$$

$$\Delta E^{\ominus} = \frac{RT}{zF} \cdot \ln K \tag{2.115}$$

Hence, the standard electrode potential difference $\Delta E^{\ominus}$ of an electrochemical cell can be used to determine the equilibrium constant K of the system.

2.3.2.4 Reference Electrodes

The electric potential of a half-cell is a potential difference itself, namely between a solid metal (electrode) and the electrolyte solution in which it is immersed. This potential difference is called the electrode potential, and it is physically impossible to measure its value. However, as we have already introduced above, it is possible to measure the difference between the electrode potential of one half-cell when combining it with another half-cell. If one chooses a particular half-cell for reasons of comparison, i.e. a ***reference cell***, then standardised electrode potentials can be measured and tabulated.

In this context, the standard hydrogen electrode has been introduced as the reference standard, whereby the electric potential of a platinum electrode which is exposed to H_2 gas with a pressure of

p_{normal} = 1.013bar(1bar = 100kPa) at θ_{normal} = 25℃ and immersed into a solution with $a(H^+)$ = 1 M is arbitrarily set to $E^\ominus = 0$:

$$H_2(g) \longrightarrow 2H^+(aq) + 2e^- \qquad E^\ominus = 0$$

In order to determine standard electrode (or redox) potentials, one thus measures the EMF of a cell, in which the concentration of solutions are all 1M. When determining the standard electrode potentials of any electrode, the standard hydrogen electrode is chosen as the left electrode (see Fig. 2.14).

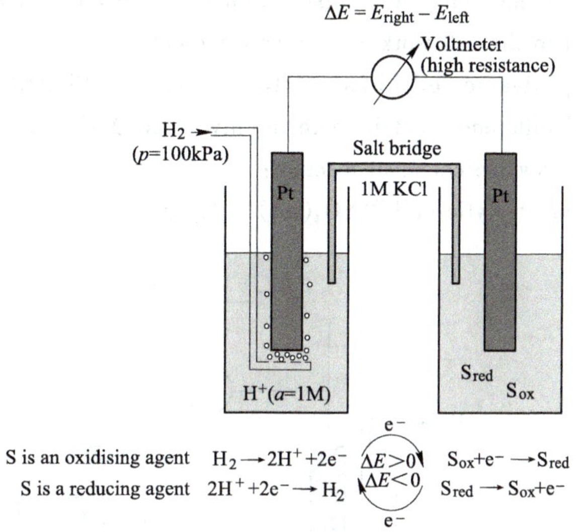

Fig. 2.14 The standard hydrogen electrode and the principle of determining standard electrode potentials

Since the standard hydrogen electrode requires a rather elaborate experimental setup, which proves impractical for many routine laboratory applications, other, more convenient reference electrodes are frequently being used. In principle, any half-cell can be employed which maintains a potential that remains practically unchanged during the course of an electrochemical measurement. One needs to evaluate the standard potential difference of the chosen electrode to the standard hydrogen electrode and correct for this difference, when deploying the chosen electrode in a measurement. The most commonly used reference electrodes in this context are:

Standard hydrogen electrode:

$$H_2(g)\,|\,H^+(aq) \qquad E^\ominus = 0 \quad 2H^+(aq) + 2e^- \longrightarrow H_2(g)$$

Silver/silver chloride electrode:

$$Ag(s)\,|\,AgCl(s) \qquad E^\ominus = 0.22V \qquad AgCl(s) + e^- \longrightarrow Ag(s) + Cl^-(aq)$$

Calomel electrode:

$$Hg_2Cl_2(s)\,|\,Hg(l) \qquad E^\ominus = 0.26V \qquad Hg_2Cl_2(s) + 2e^- \longrightarrow 2Hg(l) + 2Cl^-$$

Secondary reference electrodes such as the silver/silver chloride or the calomel electrode that are required to maintain a constant potential employ a rather insoluble metal ion salt ($AgCl$, Hg_2Cl_2) together with the elementary metal (Ag, Hg). In these cases, the potential difference with respect to the standard hydrogen electrode is not just the standard potential difference of the redox reaction. Additionally, the transition of the metal ion from the solid to the dissolved state needs to be taken into account as well (see Sect. 2.3.1.7).

2.3.2.5 External Potential Difference

When combining two half-cells into an electrochemical cell, that cell can be set up as a galvanic or electrolytic cell by choosing an appropriate external potential difference ΔE_{ext} (see Fig. 2.15). If the electric potential difference of a cell is ΔE_{cell}, then any external potential difference that is more positive will result in the cell working as a galvanic cell:

$$\Delta E_{ext} > \Delta E_{cell} \text{ galvanic cell } Pb(s) + H_2SO_4(aq) \longrightarrow PbSO_4(aq) + H_2(g)$$

An external potential difference that is more negative than ΔE_{cell} will operate the cell as an electrolytic cell with the reverse chemical reaction:

$$\Delta E_{ext} < \Delta E_{cell} \text{ electrolytic cell } PbSO_4(aq) + H_2(g) \longrightarrow Pb(s) + H_2SO_4(aq)$$

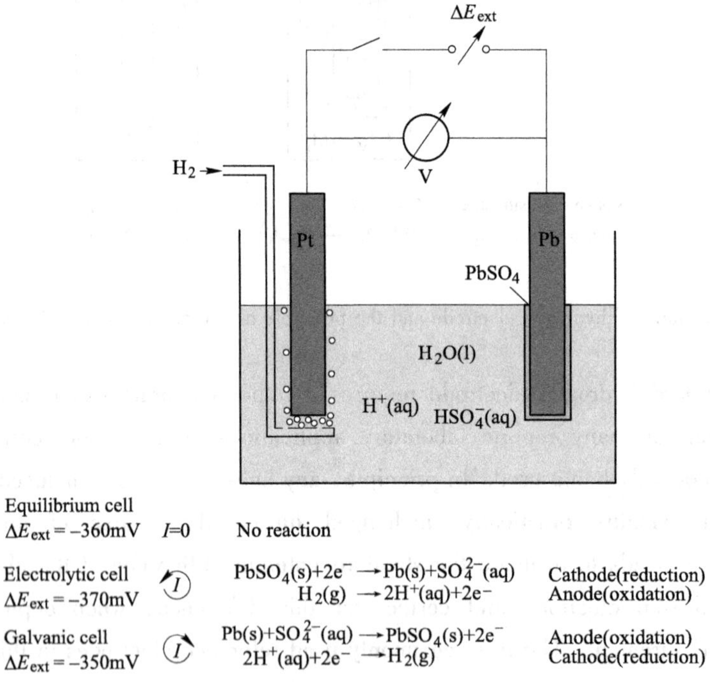

Fig. 2.15 Depending on the external potential difference, an electrochemical cell can either run as electrolytic cell, galvanic cell or rest at equilibrium

2.3.3 Nernst Equation

2.3.3.1 Equations for the EMF of Galvanic Cells

Let us write ΔG_m for the Gibbs energy change occurring during a chemical reaction m involving the conversion of v_j moles of reactants (index "react") and products ("prod"). We find that

$$\Delta G_m = \sum_{prod} v_j \mu_j - \sum_{react} v_j \mu_j \tag{2.116}$$

when the concentrations and hence the chemical potentials of the components do not change markedly during the reaction (when the degree of conversion is low). For electrode reactions involving n electrons, Eq. (2.116) can be written as

$$\pm \Delta G_{\mathrm{m}} = \sum_{\mathrm{ox}} v_j \bar{\mu}_j - \sum_{\mathrm{red}} v_j \bar{\mu}_j + n\bar{\mu}_e \tag{2.117}$$

where the plus sign is used for the anodic and the minus sign for the cathodic direction of the reaction. The electrons that are involved in the reaction are at a potential $\psi^{(\mathrm{M})}$ inside the metal, and all ions are at a potential $\psi^{(\mathrm{E})}$ inside the electrolyte. Using the expanded form for the electrochemical potential of these components and allowing for the definition of the Galvani potential, we can transform Eq. (2.116) to

$$\pm \Delta G_{\mathrm{m}} = \sum_{\mathrm{ox}} v_j \bar{\mu}_j - \sum_{\mathrm{red}} v_j \mu_j + n\mu_e - nF\phi_{\mathrm{G}} \tag{2.118}$$

It follows from this equation that the Galvani potential in the point of equilibrium of the electrode reaction ($\Delta G_{\mathrm{m}} = 0$) is given by

$$\phi_{\mathrm{G}} = \frac{1}{nF}\left(\sum_{\mathrm{ox}} v_j \mu_j - \sum_{\mathrm{red}} v_j \mu_j + n\mu_e\right) \tag{2.119}$$

None of the terms of this equation can be determined experimentally.

For a galvanic cell with two electrodes where the overall current-producing reaction takes place, allowing for charge balance we can write for the Gibbs energy change of this reaction:

$$\Delta G_{\mathrm{m}} = \sum_1 v_j \mu_j - \sum_2 v_j \mu_j \tag{2.120}$$

The reaction will occur spontaneously (without an external supply of energy) and the value of ΔG_{m} will be negative when the cell is operated as a battery [i.e., when electrode is the anode (the negative electrode)].

The OCV of this cell is the algebraic sum of the Galvani potentials at three interfaces. When each of them is in equilibrium, we find for the overall potential difference (which in this case we can call an electromotive force or EMF):

$$\xi = \sum_2 v_j \frac{\mu_j}{nF} + \frac{\mu_e(M_2)}{F} - \sum_1 \frac{v_j \mu_j}{nF} - \frac{\mu_e(M_1)}{F} + \left[\frac{\mu_e(M_2)}{F} - \frac{\mu_e(M_1)}{F}\right] \tag{2.121}$$

or, after simplification and allowing for Eq. (2.120)

$$\xi = \frac{\sum_2 v_j \mu_j - \sum_1 v_j \mu_j}{nF} = -\frac{\Delta G_{\mathrm{m}}}{nF} \tag{2.122}$$

This equation links the EMF of a galvanic cell to the Gibbs energy change of the overall current-producing reaction. It is one of the most important equations in the thermodynamics of electrochemical systems. It follows directly from the first law of thermodynamics, since $nF\xi$ is the maximum value of useful (electrical) work of the system in which the reaction considered takes place. According to the basic laws of thermodynamics, this work is equal to $-\Delta G_{\mathrm{m}}$.

The open cell discussed was considered as an equilibrium cell since equilibrium was established across each individual interface. However, the cell as a whole is not in equilibrium; the overall Gibbs energy of the full reaction is not zero, and when the circuit is closed, an electric current flows that is attended by chemical changes (i.e., a spontaneous process sets in).

It is typical that in Eq. (2.121) for the EMF, all terms containing the chemical potential of electrons in the electrodes cancel in pairs, since they are contained in the expressions for the Galvani potentials, both at the interface with the electrolyte and at the interface with the other electrode. This is due to the fact that the overall current producing reaction comprises the transfer of electrons across the interface between two metals in addition to the electrode reactions.

This gives rise to an important conclusion. For nonconsumable electrodes that are not involved in the current-producing reaction, and for which the chemical potential of the electrode material is not contained in the equation for electrode potential, the latter (in contrast to a Galvani potential) depends only on the type of reaction taking place; it does not depend on the nature of the electrode itself.

2.3.3.2 Nernst Equation for Ideal Systems

Electrode potentials (as well as values of the EMF of galvanic cells) depend on the composition of the electrolyte and other phases of variable composition. The electrode potential corresponds to the Galvani potential of the electrode-electrolyte interface, up to a constant term $E = \phi_G + \text{const}$. Introducing the concentration dependence of the chemical potential μ into Eq. (2.119), we find that

$$E = E_c^0 + \frac{RT}{nF}\left(\sum_{ox} v_j \ln c_j - \sum_{red} v_j \ln c_j\right) \qquad (2.123)$$

The partial pressures p_j will appear here for gaseous substances instead of concentrations c_j.

In Eq. (2.123), E_c^0 is a constant that depends on the nature of the electrode reaction; it is equal to the electrode potentials for values $c_j = 1$. The value of the electrode potential E depends primarily on the value of this constant, since the second term of this equation, which is the correction term for concentrations, is relatively small, although in certain cases it becomes the major term. The components that are involved in the electrode reaction and therefore influence the value of the electrode potential are called potential-determining substances. It follows from this equation that the potential will shift in the positive direction when the concentration of the oxidizing agent or of components entering the reaction together with it is raised, but it will shift in the negative direction when the concentration of the reducing agent or of components entering the reaction together with it is raised

The combination of constants RT/F often appears in electrochemical equations; it has the dimensions of voltage. At 25℃ (298.15K) it has a value of 0.02569V (or roughly 25mV). When including the conversion factor for changing natural to common logarithms, we find a value of 0.05916V (about 59 or 60mV) for 2.303 (RT/F) at 25℃. Values for other temperatures can be found by simple conversion, since this parameter is proportional to the absolute temperature

For metal electrodes of the first type at which the reaction $M^{z+} + ze^- \rightarrow M$ takes place, Eq. (2.123) becomes

$$E = E_c^0 + \frac{RT}{z + F}\ln c_{M^{z+}} \qquad (2.124)$$

An equation of this form had been suggested by Walther Nernst in 1869. In the case of amalgam electrodes (e.g., sodium amalgam in NaOH solution), the potential depends on the concentrations of the oxidizing agent (the cations in the electrolyte) as well as of the reducing agent (the metal in the amalgam, $c_{Na}^{(Hg)}$):

$$E = E_c^0 + \frac{RT}{F}\left(\ln \frac{c_{Na^+}^{(E)}}{c_{Na}^{(Hg)}}\right) \tag{2.125}$$

An equation for the very simple redox reaction $Fe^{3+} + e^- \longrightarrow Fe^{2+}$,

$$E = E_c^0 + \frac{RT}{F}\left(\ln \frac{c_{Fe}^{3+}}{c_{Fe}^{2+}}\right) \tag{2.126}$$

had been suggested by Franz C. A. Peters in 1898. At present, all equations of this type are known as ***Nernst equations***.

2.3.3.3 Equations for Real Systems

Equations describing the relation between electrode potential and composition of the system in the case of real systems can be written, by analogy with Eq. (2.123), as functions of the component activities

$$E = E_a^0 + \frac{RT}{nF}\left(\sum_{ox} v_j \ln a_j - \sum_{red} v_j \ln a_j\right) \tag{2.127}$$

For the particular case of electrodes of the first type, we have, instead of Eq. (2.124),

$$E = E_a^0 + \frac{RT}{z + F}(\ln a_{M^{z+}}) \tag{2.128}$$

There is a major difficulty that arises when such equations are used in practice, in that the activities of individual ions are unknown unless the solutions are highly dilute and the ionic components involved in the electrode reaction do not form electroneutral groups. Hence, for practical calculations we must employ values of mean ionic activity $a_\pm$:

$$E = E_a^0 + \frac{RT}{nF}\left(\sum_{ox} v_j \ln a_\pm - \sum_{red} v_j \ln a_\pm\right) \tag{2.129}$$

which makes the results somewhat provisional. But this can be tolerated insofar as the experimental values of electrode potentials are also slightly distorted and depend on the activities of other ions, because reference electrodes with other potential-determining ions or liquid-junction potentials are present. In cases where data for the activity coefficients are altogether unavailable (particularly in multicomponent solutions), concentrations must be used instead of activities [Eq. (2.123)]. This leads to appreciable errors and is admissible only for the purposes of approximate calculations.

Parameter E_a^0 in Eqs. (2.127) and (2.129) is called the standard electrode potential; it corresponds to the value of electrode potential that is found when the activities of the components are unity. Values E_a^0 differ somewhat from values E_c^0. For a more distinct differentiation between these parameters, E_c^0 is called the formal electrode potential.

The overall current-producing reaction in a cell involves only electroneutral sets of species; hence, in the equation for the EMF ξ of such cells,

$$\xi = E_+ - E_- \tag{2.130}$$

It is entirely legitimate to replace the true activities of individual ions by mean ionic activities. Subscripts a and c for the parameter E^0 are dropped in the following.

2.3.3.4 Electrodes of the Second Kind

A silver electrode in aqueous KCl solution is the example of an electrode of the second type. Here the reaction

$$Ag + Cl^- \rightleftharpoons AgCl + e^- \quad (2.131)$$

takes place. Owing to the low solubility of AgCl, the product of the anodic process precipitates as a solid and the electrolyte is almost always saturated with it. The solubility product $L(AgCl) = a_{Ag}+ a_{Cl^-} = 1.56\times10^{-10}$ mol^2/L^2; that is, the 1M KCl solution is saturated with AgCl at an Ag^+ ion concentration as low as about 10^{-10} mol/L.

Since in a saturated solution of a silver salt the activity of the Ag^+ ions depends on the activity of the anions, the potential of this electrode will depend on the activity of the anions

$$E = E^0 - \frac{RT}{F}\ln a_{Cl^-} \quad (2.132)$$

Here the value of E^0 (Ag, AgCl) differs from that of a silver electrode of the first type, E^0(Ag, Ag^+):

$$E^0(Ag, AgCl) = E^0(Ag, Ag^+) - \frac{RT}{F}\ln l(AgCl) \quad (2.133)$$

It is sometimes said that this electrode is reversible with respect to the anion. This claim must be examined in more detail. An electrode potential that depends on anion activity still constitutes no evidence that the anions are direct reactants. Two reaction mechanisms are possible at this electrode, a direct transfer of chloride ions across the interface in accordance with Eq. (2.131) or the combination of the electrode reaction $Ag \rightarrow Ag^+ + e^-$ (transfer of the Ag^+ ions) with the ionic reaction in the bulk solution: $Ag^+ + e^- \rightarrow AgCl$. The overall reaction is the same; hence, in both cases Eq. (2.131) is legitimate, yet in the second case the chloride ions are additional, not primary reactants. Thus, thermodynamic data do not suffice if we want to unravel the true mechanism of an electrode process.

Electrodes of the second type can formally be regarded as a special case of electrodes of the first type where the standard state (when $E=E^0$) corresponds not to $a_{Ag^+} = 1$ but to a value of $a_{Ag^+} \approx 10^{-10}$ mol/L, which is established in a KCl solution of unit activity. In this case, the concentration of the potential-determining cation can be varied by varying the concentration of an anion, which might be called the controlling ion. The oxides and hydroxides of most metals (other than the alkali metals) are poorly soluble in alkaline solutions; hence, almost all metal electrodes in alkaline solutions are electrodes of the second type.

A picture similar to that just described is seen for metal electrodes when the solution contains a complexing agent (e.g., for the silver electrode when KCN has been added to a $AgNO_3$ solution). Then the complex formation equilibrium

$$Ag^+ + 3CN^- \longrightarrow [Ag(CN_3)]^{2-} \quad (2.134)$$

will be established in the solution, and as a result the concentration of free Ag^+ ions decreases drastically. By suitable selection of complexing agents, one can alter the equilibrium potentials of metal electrodes within wide limits.

2.3.3.5 Nernst Equation at Very Low Concentrations

At zero concentration of the potential-determining substances, the values of electrode potential calculated with Eq. (2.123) or (2.127) tend toward $\pm\infty$, which is physically meaningless. This implies that these equations cannot be used below a certain concentration

Two types of notion exist with respect to the term low concentrations [i. e., a low absolute concentration (highly dilute solutions) and a low equilibrium concentration (as in the formation of complexes or compounds of low solubility)]. In the latter case, when potential-determining substances start to be withdrawn from the solution, they reform because of the shift in equilibrium (i. e., their potential supply is large).

The Nernst equation is of limited use at low absolute concentrations of the ions. At concentrations of 10^{-7} to 10^{-9} mol/L and the customary ratios between electrode surface area and electrolyte volume ($S/V \approx 10^{-2}$ cm^{-1}), the number of ions present in the electric double layer is comparable with that in the bulk electrolyte. Hence, EDL formation is associated with a change in bulk concentration, and the potential will no longer be the equilibrium potential with respect to the original concentration. Moreover, at these concentrations the exchange current densities are greatly reduced, and the potential is readily altered under the influence of extraneous effects. An absolute concentration of the potential-determining substances of 10^{-5} to 10^{-7} mol/L can be regarded as the limit of application of the Nernst equation. Such a limitation does not exist for low-equilibrium concentrations.

2.3.4 The Electrochemical Reaction Rate

2.3.4.1 Intermediate Reaction Steps

Electrochemical reactions in which two or more electrons are transferred more often than not follow a path involving a number of consecutive, simpler steps producing stable or unstable intermediates (i. e., they are multistep reactions).

The set of all intermediate steps is called the reaction pathway. A given reaction (involving the same reactants and products) may occur by a single pathway or by several parallel pathways. In the case of invertible reactions, the pathway followed in the reverse direction (e. g., the cathodic) may or may not coincide with that of the forward direction (in this example, the anodic). For instance, the relatively simple anodic oxidation of divalent manganese ions which in acidic solutions yields tetravalent manganese ions: $Mn^{2+} \rightarrow Mn^{4+} + 2e^-$, can follow these two pathways

$$Mn^{2+} \xrightarrow{-e^-} Mn^{3+} \xrightarrow{-e^-} Mn^{4+} \qquad (\mathrm{I})$$

$$2[Mn^{2+} \xrightarrow{-e^-} Mn^{3+}]; \qquad Mn^{3+} \longrightarrow Mn^{2+} + Mn^{4+} \qquad (\mathrm{II})$$

The second pathway includes a step in which the trivalent manganese ions formed as intermediates disproportionate.

It is convenient to represent multistep electrode reactions involving one or more pathways in the form of tables such as this:

k	Step	l_k	μ_k(Ⅰ)	μ_k(Ⅱ)
1	$Mn^{2+} \rightarrow Mn^{3+} + e^-$	1	1	2
2	$Mn^{3+} \rightarrow Mn^{4+} + e^-$	1	1	
3	$Mn^{3+} \rightarrow Mn^{2+} + Mn^{4+}$	0		1

where l_k is the number of electrons in a step, and μ_k is the stoichiometric number of a step in pathways Ⅰ and Ⅱ, which indicates how many times this step is repeated in an elementary reaction act (i. e., during the loss of two electrons from one Mn^{2+} ion). For instance, in a reaction following the second pathway, the first step occurs twice ($\mu_1 = 2$), but one of the Mn^{2+} ions that had reacted is regenerated in the step that follows.

For electrochemical steps as a rule, $l_k = 1$, although for the sake of generality we shall retain the symbol l_k in the equations. Purely chemical steps are also possible, in which electrons are not involved and $l_k = 0$ (e. g., the third step in the example above). Obviously, $\sum \mu_k l_k = n$, where n is the total number of electrons involved in one elementary reaction act (here $n = 2$). Electrochemical steps are often denoted by the letter E (or e), and chemical steps by the letter C (or c). Thus, the first pathway in the example above can be said to follow an EE scheme, and the second an EC scheme. The reactions considered below will occur by only a single pathway (in both the forward and reverse directions), and there will be no parallel path.

In multistep reactions, the number of particles of any intermediate B_k produced in unit time in one of the steps is equal to the number of particles reacting in the next step (in the steady state the concentrations of the intermediates remain unchanged). Hence, the rates of all intermediate steps are interrelated. Writing the rate v_k of an individual step as the number of elementary acts of this step that occur in unit time, and the rate v of the overall reaction as the number of elementary acts of the overall reaction that occur within the same time, we evidently have

$$v = \frac{v_1}{\mu_1} = \frac{v_2}{\mu_2} = \cdots = \frac{v_k}{\mu_k} = \cdots = \frac{v_z}{\mu_z}$$

Therefore, in the steady state the reduced rates v_k/μ_k are identical for all steps and equal to the rate of the overall reaction.

Each of the intermediate electrochemical or chemical steps is a reaction of its own (i. e., it has its own kinetic peculiarities and rules. Despite the fact that all steps occur with the same rate in the steady state, it is true that some steps occur readily, without kinetic limitations, and others, to the contrary, occur with limitations. Kinetic limitations that are present in electrochemical steps show up in the form of appreciable electrode polarization. It is a very important task of electrochemical kinetics to establish the nature and kinetic parameters of the intermediate steps as well as the way in which the kinetic parameters of the individual steps correlate with those of the overall reaction.

2.3.4.2 Rate-determining step

Consider the correlation between the kinetic parameters of the overall reaction and those of its individual steps in the instance of a very simple, invertible two-step chemical reaction:

$$A \underset{}{\overset{\pm 1}{\rightleftharpoons}} B \underset{}{\overset{\pm 2}{\rightleftharpoons}} D \tag{2.135}$$

where B is the reaction intermediate and $\mu_1 = \mu_2 = 1$. We shall assume that both steps are first order in the reactants, that is,

$$v_1 = k_1 c_A - k_{-1} c_B ;\ v_2 = k_2 c_B - k_{-2} c_D \tag{2.136}$$

In the steady state, $v_1 = v_2 = v$. With the values of v_1 and v_2 from Eq. (2.136), we find that

$$c_B = \frac{k_1 c_A + k_2 c_D}{k_{-1} + k_2} \tag{2.137}$$

for the steady concentration of intermediate B and

$$v = \frac{k_1 k_2 c_A - k_{-1} k_{-2} c_D}{k_{-1} + k_2} \tag{2.138}$$

for the rate of the overall reaction.

The direction of the reaction will depend on the relative concentrations of reactants and products; the reaction will go from left to right when $k_1 k_2 c_A > k_{-1} k_{-2} c_D$, and in the opposite direction, $v<0$, when the opposite inequality holds.

When we consider the overall reaction while disregarding the formation of intermediates, we can write its rate as

$$v = k_0 c_A - k_{-0} c_D \tag{2.139}$$

By comparing Eqs. (2.138) and (2.139), we find the connection between the kinetic parameters of the overall reaction and those of the individual steps:

$$k_0 = \frac{k_1 k_2}{k_{-1} + k_2} ;\ k_{-0} = \frac{k_{-1} k_{-2}}{k_{-1} + k_2} \tag{2.140}$$

A quantity of great importance in these equations is the ratio of parameters k_{-1} and k_2(i. e., of the constants in the expressions for the rate of reconversion of the intermediates to original reactants and for the rate of conversion of the intermediates to final products). In the particular case of $k_{-1} \ll k_2$, it follows from Eq. (2.141) that

$$v = k_1 v_A - \frac{k_{-1} k_{-2}}{k_2} c_D = k_1 c_A - k_{-1} c^0_{B(D)} \tag{2.141}$$

Here $k_{-1} c^0_{B(D)}$ is the concentration of intermediate B that would be found under conditions of complete equilibrium between it and the reaction products (i. e., $k_2 c^0_{B(D)} = k_{-2} c_D$):

$$c^0_{B(D)} = \frac{k_{-2}}{k_2} c_D \tag{2.142}$$

It is a very important conclusion following from Eq. (2.141) that in the case considered, the rate of the overall reaction is determined wholly by the kinetic parameters of the first step (k_1 and

k_{-1}), while the second step influences this rate only through the equilibrium concentration of the intermediate B. We say, therefore, that the first step (with its low value of parameter k_{-2}) is ***the rate-determining step*** (RDS) of this reaction. Sometimes the term slow step is used, but this term is not very fortunate, inasmuch as the effective rates, v_1 and v_2, of the two steps actually are identical. Analogously, when $k_{-1} \gg k_2$, we have

$$v = \frac{k_1 k_2}{k_{-1}} c_A - k_{-2} c_D = k_2 c^0_{B(D)} - k_{-2} c_D \tag{2.143}$$

where

$$c^0_{B(D)} = \frac{k_1}{k_{-1}} c_A \tag{2.144}$$

In this case, the overall reaction rate is determined by the parameters of the second step (i. e., this step is now rate determining, and the concentration of the intermediate B is determined by the equilibrium of the first step).

Generally, reactant A and product D will not be in chemical equilibrium since their concentrations, c_A and c_D, are defined arbitrarily. Hence, $c^0_{B(A)}$ and $c^0_{B(D)}$ will have different values; they will coincide only in the particular case of overall equilibrium between substances A and D, which will be established at concentration ratios $c_D/c_A = k_1 k_2 / k_{-1} k_{-2}$.

It is important to note that it is precisely the ratio of k_{-1} and k_2 that decides which step is rate determining, not the ratios of the parameters of both steps in the forward reaction (k_1 and k_2) or in the reverse reaction (k_{-1} and k_{-2}).

In chemical reactions, the kinetic parameters k_k and k_{-k} are constant for given conditions (of temperature, etc.). Hence, the same step will be rate determining in the forward and reverse directions of the reaction (provided that the reaction pathways are the same in both directions).

The assumption had been made in deriving Eq. (2.137) that the concentration of the intermediate B is determined solely by the balance of rates of individual steps of the process. It is implied here that this intermediate cannot escape from the reaction zone by processes such as diffusion and evaporation.

2.3.4.3 Two-step Electrochemical Reaction

Equilibrium Conditions

All the relations reported above are valid for simple two-step electrochemical reactions, when instead of rate constants k_k of the individual steps or of the reaction as a whole, we use the corresponding kinetic parameters h_k. We shall assume for the sake of definition that in the electrochemical reaction substance A is the reducing agent and substance D is the oxidizing agent [i. e., that in Eq. (2.135) the anodic reaction is that going from left to right].

Electrochemical reactions differ fundamentally from chemical reactions in that the kinetic parameters h_k are not constant (i. e., they are not rate "constants") but depend on the electrode potential. This dependence has an important consequence: At given arbitrary values of the concentrations c_A and c_D, an equilibrium potential E_0 exists in the case of electrochemical reactions which is the potential at which substances A and D are in equilibrium with each other. At this point (E_0) the intermediate B is in common equilibrium with substances A and D. For this equilibrium concentration we obtain from Eqs. (2.143) and (2.144),

$$c_B^0 = \frac{h_1^0}{h_{-1}^0} c_A = \frac{h_{-2}^0}{h_2^0} c_D \qquad (2.145)$$

At the point of equilibrium, the exchange rate of the reaction as a whole is given by

$$v_1^0 = h_0^0 c_A = h_{-0}^0 c_B^0 \; ; \; v_2^0 = h_2^0 c_B = h_{0-2}^0 c_D \qquad (2.146)$$

and the exchange rates of the individual steps are given by

$$v_1^0 = h_0^0 c_A = h_{-0}^0 c_D \; ; \; v_2^0 = h_2^0 c_B^0 = h_{-2}^0 c_D \qquad (2.147)$$

With Eq. (2.147), this yields the following relation between the exchange rates:

$$\frac{1}{v_o} = \frac{1}{v_1^0} + \frac{1}{v_2^0} \qquad (2.148)$$

which is valid for electrochemical reactions at the equilibrium potential.

When the rate of the overall reaction is stated in electrical units [i. e., in terms of the current density (CD) $i \equiv nFv$], it will be convenient to use the concept of partial current densities of the first and second steps, which are defined as $i_1 \equiv l_1 F v_1$ and $i_2 \equiv l_2 F v_2$. In the steady state, $v = v_1 = v_2$ and $i = i_1 = i_2$. With these parameters, Eq. (2.149) becomes.

$$\frac{1}{i_o} = \frac{l_1}{i_1^0} + \frac{l_2}{i_2^0} \qquad (2.149)$$

General Kinetic Equation

The rates of an electrochemical reaction at potentials away from the equilibrium value are given by Eq. (2.138), which in this case can be written as

$$i = nF\left(h_2 c_A - \frac{h_{-1} h_{-2} c_D}{h_{-2} + h_2}\right) \qquad (2.150)$$

Since parameters h_k depend on potential, the RDS may not be the same in different regions of potential. Consider the two pairs of inequalities

$$(1a)\ h_1 h_2 c_A > h_{-1} h_{-2} c_D \quad \text{and} \quad (1b)\ h_1 h_2 c_A < h_{-1} h_{-2} c_D$$
$$(2a)\ h_2 > h_{-1} \quad \text{and} \quad (2b)\ h_2 < h_{-1} \qquad (2.151)$$

Inequality (1a) is valid for anodic polarization, inequality (1b) for cathodic polarization $i < 0$. In this case, the point of changeover is evidently the equilibrium potential E_0. However, the changeover from inequality (2a) to inequality (2b) generally occurs at another value of potential, that of the change in mechanism, E_{cm}. At potentials more positive than E_{cm} inequality (2a) holds, and step 1 is the RDS at potentials more negative than E_{cm}, step 2 is the RDS.

At high anodic potentials when the electrode potential is more positive than both E_0 and E_{cm}, inequalities (1a) and (2a) hold and step 1 is the RDS. The kinetic equation then is

$$i = nFh_1 c_A = nFk_1 c_A \exp\left(\frac{\beta_1 FE}{RT}\right) \qquad (2.152a)$$

(the reverse reaction can be disregarded). In exactly the same way, at high values of cathodic polarization when the electrode potential is more negative than both E_0 and E_{cm}, step 2 is the RDS; it is the first step in the cathodic direction [i. e., of reaction (2.135) occurring from right to left]. The kinetic equation now is

$$i = nFh_{-2}c_D = nFk_{-2}c_D\exp\left(-\frac{\beta_2 FE}{RT}\right) \quad (2.152b)$$

The behavior in the regions of moderate anodic or cathodic polarization depends on the relative positions of potentials E_{cm} and E_0, which in turn depend on the relative values of constants k_1 and k_{-2}. For E_{cm} which are more positive than E_0(Fig. 2.16(a)), relation (2.152) for the cathodic CD remains valid at all values of cathodic polarization (except for the region of low values where the reverse reaction must be taken into account). At moderate values of anodic polarization, inequalities (1a) and (2b) are found to be valid at potentials more negative than E_{cm}, while step 2 becomes rate determining, which is the second step along the reaction path. In this case [see Eq. (2.153)], we have

$$i = nF\frac{h_1}{h_{-1}}h_2c_A \quad (2.153)$$

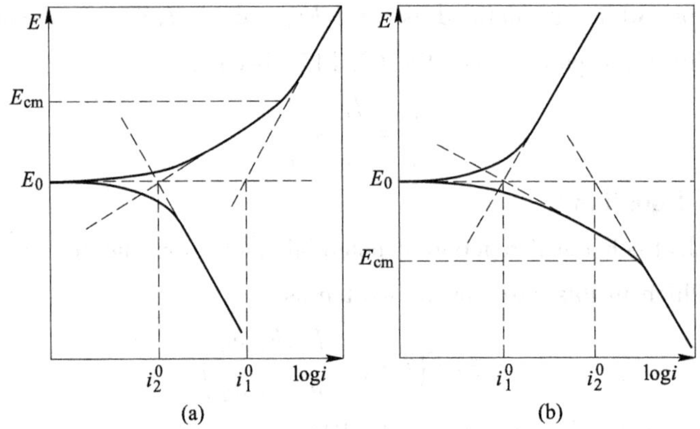

Fig. 2.16 Polarization curves for a two-step reaction

However, for E_{cm} which are more negative than E_0(Fig. 2.16(b)), relation (2.152) remains valid at all values of anodic polarization. At moderate values of cathodic polarization or potentials more positive than E_{cm}, step 1 will be rate determining, which is the second step along the reaction path, and we find for the kinetic equation

$$i = nF\frac{h_{-2}}{h_2}h_{-1}c_D = nF\frac{k_{-2}}{k_2}k_{-1}c_D\exp\left(\frac{l_2\beta_{-2}FE}{RT}\right) \quad (2.154)$$

We can see here that a formal change occurs in the polarization equation at potential E_{cm}; in particular, transfer coefficient β_0 changes. When the two steps are one electron steps l_{k1} and all transfer coefficients of the individual steps are close to 0.5, the value of β_0 will change from 0.5 in the region of very large polarization to 1.5 at lower values of polarization, on the other side of E_{cm}. Slope b in the Tafel equation changes accordingly from 0.12 to 0.04V, and the polarization curve exhibits two logarithmic sections (two Tafel slopes). Such breaks in the polarization curve (when plotted semilogarithmically) are a typical indication for multistep electrochemical reactions involving a change in the RDS.

It follows from these kinetic equations that in reactions where the RDS occurs after another step, which is an equilibrium step, the kinetic coefficients k_0 and β_0 of the overall reaction are different from the corresponding coefficients of the RDS; in fact, in the first case, $k_0 \equiv (k_1/k_{-1})k_2$ and $\beta_0 \equiv l_1+\beta_2$, and in the second case, $k_{-0} \equiv (k_{-2}/k_2)k_{-1}$ and $\beta_0 \equiv l_2+\beta_{-1}$. It is important to note that if the preceding equilibrium step is an electrochemical step ($l_1 \geqslant 1$), the transfer coefficient β_0 of the overall reaction will always be larger than unity. The sum of transfer coefficients in the forward and reverse directions of the overall reaction is given by

$$\beta_0 + \beta_{-0} = l_1 + \beta_2 + \beta_{-2} = l_1 + l_2 = n \tag{2.155}$$

However, when different steps are rate determining under cathodic and anodic polarization, the sum of transfer coefficients of these steps, $\beta_1+\beta_{-2}$ or $\beta_2+\beta_{-1}$, generally will be different from n, l_1, or l_2.

When the logarithmic sections of the polarization curves are extrapolated to the equilibrium potential E_0, they yield intercepts on the horizontal axis which represent the logarithms of exchange CD of the individual steps (up to factors n/l). By extrapolation of the section at high anodic potentials, we obtain, according to Eq. (2.152), the exchange CD $(n/l_1)i_1^0$ of step 1, but by extrapolation of the section at high cathodic potentials we obtain, again according to Eq. (2.152), the exchange CD $(n/l_2)i_2^0$ of step 2. It follows that in both cases we obtain the exchange CD of the step that comes first in the reaction path. Analogously, when extrapolating the section in the region of less important polarization (between potentials E_{cm} and E_0), we obtain, according to Eqs. (2.153) or (2.155), the exchange CD of the step that comes second in the reaction path (see Fig. 2.16).

Thus, in the region of very high anodic or cathodic polarization, the RDS is always the first step in the reaction path. The transfer coefficient of the full reaction which is equal to that of this step is always smaller than unity (for a one-electron RDS), while slope i in the Tafel equation is always larger than 0.06V. When the potential E_{cm} is outside the region of low polarization, a section will appear in the polarization curve at intermediate values of anodic or cathodic polarization where the transfer coefficient is larger than unity and b is smaller than 0.06V. This indicates that in this region the step that is second in the reaction path is rate determining.

A break in the polarization curve will not be observed when the kinetic parameters of the two steps (h_1 and E_{cm-2}, i_0^1 and i_0^2) are drastically different, and hence, potential E_{cm} is in the region of excessive anodic or cathodic polarization where measurements become impossible or where the behavior observed is distorted by other phenomena (e.g., concentration polarization).

Questions

1. Calculate the work done per mole when an ideal gas is expanded reversibly by a factor of three at 120℃.
2. A friend thinking of investing in a company that makes engines shows you the company's

prospectus. Analysing the machine, you realise that it works by harnessing the expansion of a gas. The operating temperature is approx. Constant at 120℃, and the gas doubles in volume during the power extraction phase of operation. It is claimed that the machine produces 5.5kJ/mol of work during expansion. What advice would you give your friend? Briefly explain the thermodynamic rationale.

3. Consider the gas-phase reaction

$$H_2 + I_2 \longrightarrow 2HI$$

(a) Assume that the reaction order is as suggested by the chemical equation. Calculate the rate constant at 681K, assuming that from an initial pressure of iodine of $823N/m^2$, the rate of loss of iodine was $0.192N/(m^2 \cdot s)$. The initial pressure of hydrogen was $10500N/m^2$.

(b) What is the rate of the reaction if the iodine pressure was unchanged and the initial hydrogen pressure was $39500N/m^2$?

4. The rate constant of a first-order reaction was measured as $1.11\times10^{-3}s^{-1}$.

(a) What is the half-life of the reaction?

(b) What time is needed for the concentration of the reactant to fall to 1/8 of its initial value?

(c) What time is needed for the concentration of the reactant to fall to 3/4 of its initial value?

5. Calculate the standard e. m. f. and the equilibrium constant under standard conditions for the following cell:

$$Pt(s), H_2(g) | HCl(aq) | AgCl(s), Ag(s)$$

6. Calculate the pH of a solution with a formal concentration of 5×10^{-7}M of the strong acid HI at 25℃.

7. What e. m. f. would be generated by the the following cell at 25℃, assuming ideal behaviour:

$$Pt(s), H_2(g)(1\ bar) | H^+(aq)(0.03M) | Cl^-(aq)(0.004M) | AgCl(s), Ag(s)$$

References

[1] Yoo H I. Lectures on kinetic processes in materials[M]. Cham:Springer,2020.
[2] Chen P. Material science and engineering[M]. New Brunswick:Taylor and Francis, 2013.
[3] Bagotsky S. Fundamentals of electrochemistry[M]. Hoboken:John Wiley & Sons, Inc, 2006.
[4] Hofmann A. Physical chemistry essentials[M]. Cham:Springer,2016.
[5] Smartl E, Moore E A. Solid state chemistry[M]. New Brunswick:Taylor and Francis, 2005.

Chapter 3 Foundation of Polymer Materials

3.1 Introduction of Polymer

Take a minute to look at the room and furnishings around you. Virtually everything you will see is at least partially comprised of organic-based building blocks. From the plastics that surround electronic components to individual carpet fibers, no other type of material is as heavily utilized in our society as organic-based polymers. As first defined by Staudinger in 1920, a polymer is any material that is comprised of an extended structure of small chemical repeat units, known as monomers. For simplicity, the monomeric unit is almost always clearly identified within the polymer name (i. e, poly(monomeric unit), Fig. 3. 1). A polymer is generally comprised of more than 100 repeat units; structures with lower numbers of chemical repeat units are known as oligomers. Strictly speaking, all solid-state materials with an infinite structural array are classified as polymers even inorganic structures such as metals, ceramics, and glasses. However, since we have described inorganic-based materials in previous chapters, we will focus our present discussion on polymeric materials that feature a carbon-containing backbone.

Organic-based materials are generally associated with "soft" characteristics i. e. , relatively low-melting and facile plastic deformation. Though this is generally the case for popular commercial polymers such as plastics and rubbers, there are numerous other polymer classes that exhibit hardness and thermal stabilities that even rival inorganic ceramics. Polymeric materials have long been used in packaging and other consumer product applications; however, recent developments have extended the utility range to now include microelectronics, photovoltaics, self-healing materials, and drug-delivery agents that were not envisioned the current volume of polymers used for commercial and industrial applications already exceeds the volume of steel and aluminum combined. In our era of soaring gas prices, automobile manufacturers are also scrambling to add more plastics to their vehicles. In addition to fabrication cost savings and enhanced design flexibility relative to steel, it is estimated that every 10% of weight reduction will yield a 5% increase in fuel economy. This chapter will delve into various polymer classes, with a detailed examination of the structural influence on resultant properties.

The term polymer comes from "poly" meaning many, and "mer" meaning units. Hence, polymers are composed of many units—in this case, structural units called monomers. A monomer is any unit that can be converted into a polymer. Similarly, a dimer is a combination of two monomers; a trimer is a combination of three monomers, and so on. Before describing the chemical composition of typical monomers and how they are put together to form polymers, it is useful to have a brief organic chemistry review. You may wish to refer to an organic chemistry text for more detailed information. We will reserve discussion of how these organic molecules are brought together to form polymers in this chapter when the kinetics of polymerization are described.

For the remainder of the description of polymer structure, it is sufficient to know that polymer chains are formed from the reaction of many monomers to form longchain hydrocarbons, sometimes called macromolecules, but more commonly referred to as polymers.

Molecular Weight We have described some of the general characteristics of polymers, and how they can be grouped according to structure, but we have not addressed any of the more quantitative aspects of polymer structures. For instance, we have stated that a polymer is made up of many monomer (repeat) units, but how many of these repeat units do we typically find in a polymer? Do all polymer chains have the same number of repeat units? These topics are addressed in this section on polymer molecular weight.

Degree of Polymerization The number of repeat units in an isolated polymer chain is called the degree of polymerization, $\overline{X_n}$. If the monomer from which our polymer is formed has an initial molecular weight of M_0, then the molecular weight of an isolated polymer chain is $\overline{X_n} \cdot M_0$. For example, polyethylene, $-\!\!\left(H_2C{-}CH_2\right)\!\!-$, is made up of ethylene repeat units, each of which has a molecular weight $M_0 = 2\times12 + 4\times1 = 28$g/mol. Let us assume that an isolated polyethylene chain has 1000 repeat units, so $\overline{X_n} = 1000$. The molecular weight of this isolated polyethylene chain would be 1000×(28g/mol) = 28000g/mol. Notice that we are neglecting the contribution of the terminal hydrogen located at both ends of the polyethylene chain. Since typical molecular weights are of the order 10^6 and larger, terminal groups are not a significant contribution to the molecular weight, and neglecting them does not introduce substantial error in calculations. We will also see that the molecular weight of the repeat unit in the polymer chain is not necessarily exactly the same as the molecular eight of the monomer. Small molecules can sometimes be produced during polymerization reactions, such as condensation reactions in which water is formed as a by-product. Be sure to use the molecular weight of the repeat unit in your calculations.

Average Molecular Weight In a typical polymer, not all chains have the same length. Consequently, there is a distribution of $\overline{X_n}$ values, and the molecular weight, M_i, varies from chain to chain. We could construct a histogram of the distribution of the number of chains, N_i, with molecular weight M_i, as shown in Fig. 3. 1. Molecular weights have discrete values, of course, and it is possible to find some mathematical function, $f(M_i)$ that appropriately describes the curve represented by the dotted line in Fig. 3. 1. For molecular weights, this function is simply the number fraction, n_i:

$$n_i = \frac{N_i}{\Sigma N_i} \tag{3.1}$$

Recall from calculus (or physics or statistics) that distributions can have k different moments about the origin, μ'_k, which for a discrete random variable, M_i, are given by

$$\mu'_k = \sum_i n_i (M_i)^k \tag{3.2}$$

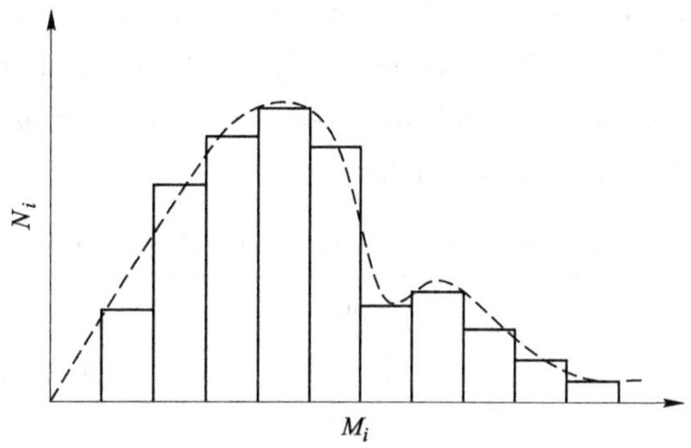

Fig. 3.1 Histogram showing the number of molecules, N_i, having the molecular weight M_i

The first moment of the distribution ($k=1$) is simply the average molecular weight, called the *number-average molecular weight*, $\overline{M_n}$,

$$\mu_1' = \overline{M_n} = \sum n_i M_i = \frac{\sum N_i M_i}{\sum N_i} \quad (3.3)$$

or, in terms of the *average degree of polymerization*, $\overline{X_n}$,

$$\overline{M_n} = \overline{X_n} M_0 \quad (3.4)$$

where M_0 is the molecular weight of a repeat unit.

The second moment of the molecular weight distribution is then

$$\mu_2' = \sum n_i M_i^2 \quad (3.5)$$

A ratio of the second moment to the first moment (number-average molecular weight) is called the *weight-average molecular weight*, $\overline{M_w}$, which is the average based on the *weight* of each polymer chain:

$$M_w = \frac{\sum N_i M_i^2}{\sum N_i M_i} = \frac{\sum n_i M_i^2}{\sum n_i M_i} \quad (3.6)$$

or, in terms of the weight of each chain, m_i (in grams),

$$\overline{M_w} = \frac{\sum m_i M_i}{\sum m_i} = \sum W_i M_i \quad (3.7)$$

where W_i is the *weight fraction*, $\frac{m_i}{\sum m_i}$. There are other types of molecular weights based on higher-order moments to the distribution (see Table 3.1), but $\overline{M_n}$ and $\overline{M_w}$ are the most widely used quantities. In general, $\overline{M_n} \neq \overline{M_w}$, but taken together as a ratio, they provide a measure of the breadth of the molecular weight distribution. The relationship between $\overline{M_n}$ and $\overline{M_w}$ can be shown in Fig. 3.2. This ratio is called the *polydispersity index*, and it takes on values greater than or equal to 1.0:

$$\frac{\overline{M_w}}{\overline{M_n}} \geqslant 1.0 \quad (3.8)$$

If all chains have exactly the same weight and number of repeat units, then the system is termed *monodispersed*, and the polydispersity index is exactly 1.0. Most real polymers have rather large polydispersity indexes, but some standards used for chromatography calibration, such as polystyrene, can have values approaching unity.

Table 3.1 Summary of the most common molecular weight averages

Average	Definition	Methods of determination
$\overline{M_n}$	$\frac{\sum N_i M_i}{\sum N_i}$	Osmotic pressure, colligative properties, end group analysis
$\overline{M_w}$	$\frac{\sum N_i M_i^2}{\sum N_i M_i}$	Light scattering, sedimentation velocity
$\overline{M_z}$	$\frac{\sum N_i M_i^3}{\sum N_i M_i^2}$	Sedimentation equilibrium
$\overline{M_v}$	$\left(\frac{\sum N_i M_i^{1+a}}{\sum N_i M_i}\right)^{\frac{1}{a}}$	Intrinsic viscosity (a is characteristic of a system and lies between 0.5 and 1.0)

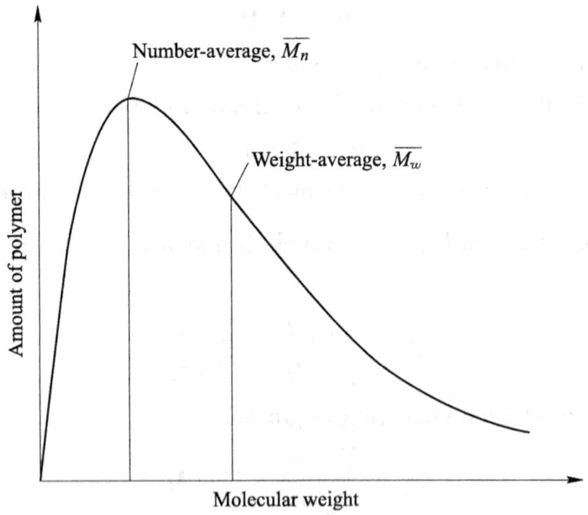

Fig. 3.2 Distribution of molecular weight for a typical polymer

Tacticity An important phenomenon in some, but not all, industrially significant polymers arises from the ability of organic monomers with asymmetric carbons to form stereoisomers. When the repeat units (monomer units) are placed together in a long chain, monomers such as the one shown in Fig. 3.3 can add to the growing chain in one of two ways: with the R group (which may be a chlorine, methyl group, or some other functional group) sticking out of the plane, or with it pointing back into the plane of the paper. The variation in how the functional groups are arranged leads to differences in tacticity.

Tacticity is variations in the configuration (not conformation!) of polymer chains as a result of the sequence of asymmetric carbon centers in the repeat units. There are three types of tacticity in polymers: isotactic, syndiotactic, and atactic. In isotactic carbon, as in Fig. 3.4(a). Fig. 3.4(b) shows the syndiotactic arrangement, in which the R groups perfectly alternate between sticking in and out of the plane of the carbon backbone. Finally, it is possible (and common) to have a perfectly random arrangement of functional groups, as in Fig. 3.4(c), in which case an atactic polymer results. Notable examples of polymers that exhibit tacticity are polyvinylchloride (PVC, R = Cl), polystyrene (PS, R = Ph, phenyl group), and polypropylene (PP, R = CH_3).

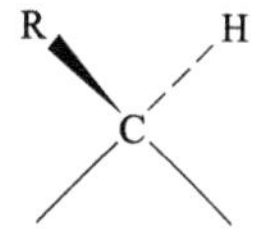

Fig. 3.3 An example of an asymmetric carbon atom

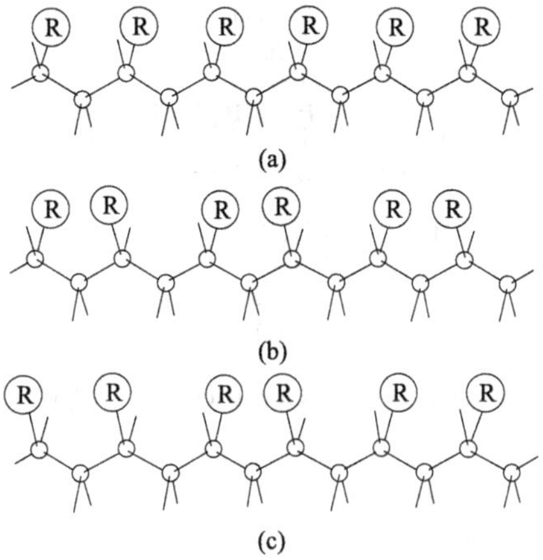

Fig. 3.4 Summary of different types of tacticity, where R represents an atom other than hydrogen
(a) Isotactic; (b) Syndiotactic; (c) Atactic

Homopolymers and Copolymers Fig. 3.5 shows a long chain which are composed of structural entities called mer units, which are successively repeated along the chain. "Mer" originates from the Greek word meros, which means part; the term polymer was coined to mean many mers. We sometimes use the term monomer, which refers to a stable molecule from which a polymer is synthesized. When all the repeating units along a chain are of the same type, the resulting polymer is called a homopolymer. There is no restriction in polymer synthesis that prevents the formation of compounds other than homopolymers.

Polymers can be made that contain more than one type of repeat unit. For example, the R group on the asymmetric carbon in Fig. 3.3 could be chlorine in some of the monomer units and fluorine in the rest. Such polymers are called *copolymers*. The ratio of the two types of monomers can vary from 0 to 1, and there can be more than two types of monomers in a copolymer. The presence of more than one type of repeat unit opens up many possibilities for variation in the structure of the polymer, or *chain architecture* as it is sometimes called. For example, the following reaction is made by the author (see Fig. 3.6).

We will not describe the myriad of possible variations and the important consequences in terms of polymer physical properties, but here merely categorize copolymers in some broad, structural

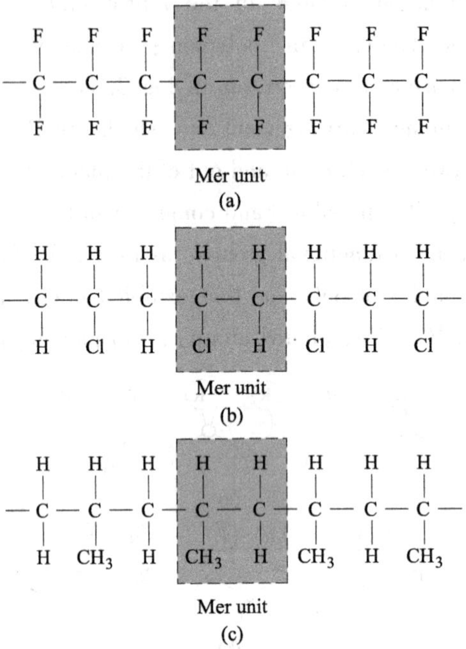

Fig. 3. 5 Mer and chain structures for polytetrafluoroethylene(a), polyvinyl chloride(b), polypropylene(c)

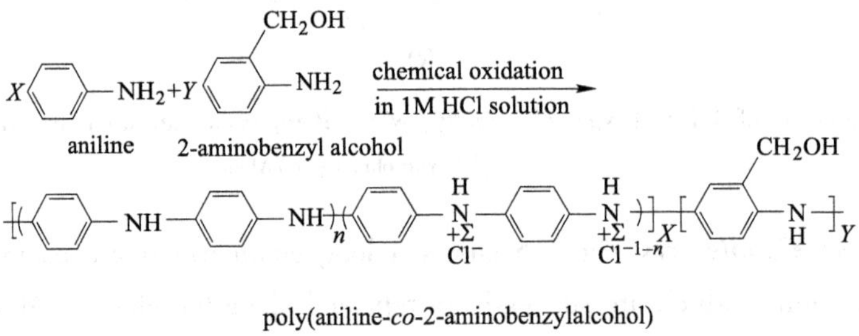

Fig. 3. 6 Preparation of poly(aniline-*co*-2-aminobenzyl alcohol)

terms. If the two monomers of a bi-component copolymer are perfectly alternating, an *alternating copolymer* results, as shown in Fig. 3. 7(a). If the monomer units alternate in a random fashion, a *random copolymer* results, as in Fig. 3. 7(b). Fig. 3. 7(c) shows a *block copolymer*, in which the different monomers prefer to add to each other, resulting in large segments of one type of monomer, followed by large segments of the other type of monomer. Finally, a second type of monomer can be added as a branch to a polymer backbone, as shown in Fig. 3. 7(d), resulting in a *graft copolymer*. These different types of copolymers can have markedly different physical properties and are the basis for many important materials in the healthcare and consumer products industries.

Thermoplastic Polymers Most thermoplastic polymers are used in high volume, widely recognized applications, so they are often referred to as *commodity plastics*. (We will elaborate upon the distinction between a polymer and a plastic in Chapter 7, but for now we simply note that a plastic

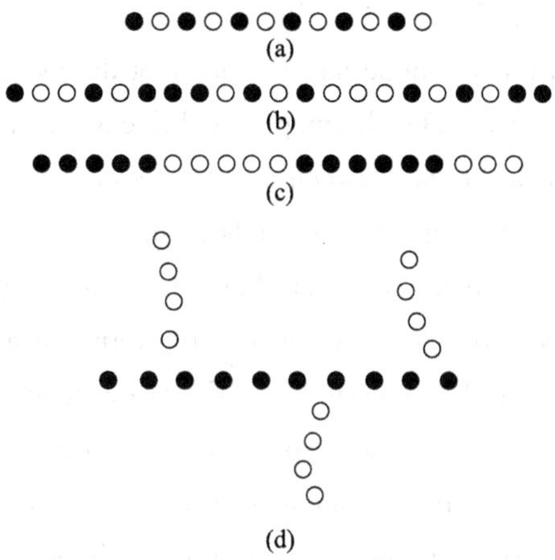

Fig. 3.7 Summary of copolymer classifications
(a) Alternating; (b) Random; (c) Block; (d) Graft

is a polymer that contains other additives and is usually identified by a variety of commercial trade names. There are numerous databases, both in books and on the Internet, that can be used to identify the primary polymer components of most plastics. With a few notable exceptions, we will refer to most polymers by their generic chemical name.) The most common commodity thermoplastics are polyethylene (PE), polypropylene (PP), polyvinyl chloride (PVC) and polystyrene (PS). These thermoplastics all have in common the general repeat unit $-(CHX—CH_2)-$, where—X is—H for PE, $—CH_3$ for PP, —Cl for PVC, and a benzene ring for PS. When we discuss polymerization reactions in Chapter 3, we will see that all of these thermoplastics can be produced by the same type of reaction.

In their simplest forms, the thermoplastics are linear, carbon chain polymers. There are methods for creating branches, especially in polyethylene, while still maintaining thermoplasticity. Increased branching tends to decrease the density, melting point, and certain mechanical properties of the polymer, but increases transparency and impact toughness. Thus, branched PE is important for many packaging applications. Other special types of PE include low-density PE (LDPE), high-density PE (HDPE), and linear low-density PE (LLDPE). HDPE is used when greater stiffness is required, such as in milk, water and detergent bottles. LDPE is used for many standard piping applications. LLDPE has a low density like LDPE, but a linear structure much like HDPE. It is less expensive to produce than LDPE, and it generally has better mechanical properties than LDPE. PP, PVC, and PS have in common an asymmetric carbon atom in their backbone, a fact that leads to interesting structural properties and is elaborated upon in the next section. They all have specific advantages leading to a variety of applications, which will also be discussed in subsequent chapters. For the moment, it is important to recognize that they are linear thermoplastics, whose properties can be tailored through blending, branching, and additives. It is important to note at this point that the ability of thermoplastics to soften when heated and harden upon cooling is what leads to the principle of *recycling*. The structure of some of these common thermoplastics can be found in Appendix 2.

Thermoset Polymers There are many important examples of thermoset polymers, a subset of which are sometimes referred to as *resins*. The thermoplastic PE can be treated by electron radiation or chemical means to form chemical bonds between adjacent chains called *crosslinks*. In this process, some of the carbon-hydrogen or carbon-carbon bonds in the linear chain are broken, creating free radicals, which react with free radicals on other chains to form bridges between the chains. When the crosslinking is brought about by chemical means, the term *curing* is often employed.

The result is that unlike linear PE, crosslinked PE has no distinct melting point due to limited chain obility and will eventually degrade upon heating. This is a common characteristic of thermoset polymers. Due to their three-dimensional structure, they cannot be reshaped upon heating. Instead, they tend to remain rigid, or soften only slightly, until the backbone begins to break, leading to *polymer degradation*. While this may not seem like a desirable quality, it means that thermoset polymers are generally stiffer than thermoplastics. In fact, the crosslinks tend to increase the long-term thermal and dimensional stability of the polymer, such that thermosets find wide use in very large and complex parts where thermal stability is important. The chemical structures of thermosets are generally much more diverse than the commodity thermoplastics. The most common types of thermosets are the phenol formaldehydes (PF), urea-formaldehydes (UF), melamine-formaldehydes (MF), epoxies (EP), polyurethanes (PU), and polyimides (PI). Appendix 2 shows the chemical structure of these important thermosetting polymers.

A related, yet distinctly different, class of crosslinked polymers is the *elastomer*. Though they are structurally different than the thermosets, we will include them here since they tend to decompose when heated, rather than flow. The presence of crosslinks in some polymers allows them to be stretched, or elongated, by large amounts. Polymers that have more than 200% elastic elongation (three times the original length) and can be returned to their original form are termed elastomers. They are like thermoplastics in that they readily elongate, but the presence of crosslinks limits the elongation prior to breakage and allow the polymer to return to its original shape. Natural rubber is an important elastomer. Crosslinks are added to an emulsion of rubber, called *latex*, through the addition of heat and sulfur. The sulfur creates chemical bonds between the rubber chains in a process known as *vulcanization*. Other common elastomers are polyisoprene, butadiene rubber (BR), styrene butadiene rubber (SBR), silicones, and fluoroelastomers.

3.2 Classification of Polymers

Polymer is actually a general name given to a variety number of materials with macromolecular structures. These materials exist in countless forms and numbers because of a very large number and types of atoms present in their molecular chains, Polymers with different chemical structures will show different physical and mechanical behavior. Therefore, it is necessary to classify polymers based on their source, application, chain structure, thermal behavior, and polymerization.

3.2.1 Classification by Source

Depending on their source or origin, polymers can be classified as natural polymers, synthetic polymers and biopolymers.

3.2.1.1 Natural Polymers

Those isolated from natural materials are called natural polymers. The typical natural polymers include cotton, silk, wool, lignin, and natural rubber. In practice, some of natural polymers have experienced chemical modification in order to have more desirable properties. They are called natural modified polymers. The examples are casein, cellulose, gelatin, etc.

3.2.1.2 Synthetic Polymers

Polymers obtained through synthetic reactions with low molecular weight compounds called synthetic polymers. Typical examples are polyethylene (PE), polyvinyl chloride (PVC), nylon, and polyester. The original sources of this group of polymers are petroleum, coal and agricultural products. Synthetic polymers are the most important group of polymer materials, because most polymeric materials used widely today belong to them.

For synthetic polymers, not all the molecular chains have the same length, i.e. the molecular weight of synthetic polymers is polydisperse. Consequently, by the strict sense, such polymers are not pure compounds but mixtures of polymer chains with different chains with different chain lengths.

3.2.1.3 Biopolymers

This case involves substances like proteins, polysaccharides (cellulose, starch), polyamino acid and several gums. The molecules of proteins contain many types of monomers (up to 20 aminoacids), linked in a specific sequence (according to the genetic code) to chains of definite length. So these biopolymer molecules are identical, the polymer is monodisperse, and represents a single, pure compound.

3.2.2 Classification by Application

According to their ultimate forms and applications, polymers can be classified as plastics, elastomers or rubbers, fibers, coatings and adhesives. When a polymer is shaped into hard and tough utility articles by exerting heat and pressure, it is used as a "plastic". Some very important plastics are PE, PP, PS, PVC, PMMA Elastomers are those polymers which have flexible chains and exhibit long elongation and elasticity. Typical examples are natural rubber (polyisoprene), polybutadiene, polychlorobutadiene. As fibers we already know nylon, polyester, polyacrylonitrile, etc. In the field of coatings and adhesives, some linear and crosslinked polymers will be covered, they are polyvinyl acetate, polyacrylates, epoxy resin, etc.

3.2.3 Classification by Chain Structure

Polymers can be classified as linear, branched, or crosslinked polymers depending on to their structure. In the previous discussion on the different types of polymers and polymerizations, we have considered only those polymers in which the monomer molecules have been linked together in one continuous length to form the polymer molecule. Such polymers are termed linear polymers. Under certain reaction conditions or with certain kinds of monomers, polymers would be generated with more than two chain ends per molecule, which are called branched polymers. For branched polymer molecules, there are side branches of linked monomer units protruding from various central branch points along the main polymer chain. The difference between the shapes of linear and branched polymer molecules can be seen from the structural representations in Fig. 3. 8.

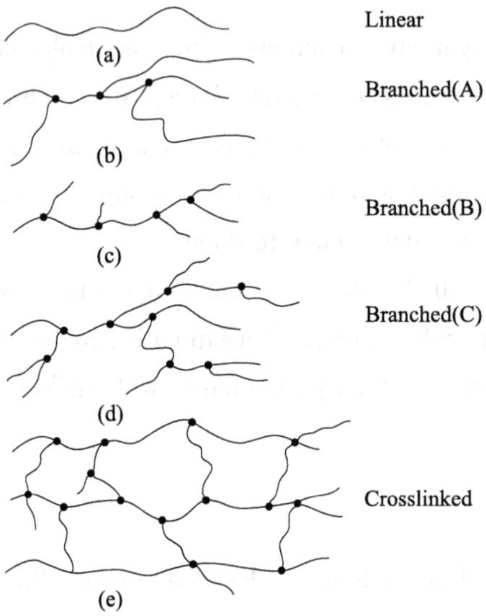

Fig. 3. 8 Structure of linear, branched, and crosslinked polymers
(a) Linear; (b) Branched(A); (c) Branched(B); (d) Branched(C); (e) Crosslinked

The branch points are indicated by heavy dots. The illustrations show that there are several different kinds of branched polymers. The branched polymer can be comb like in structure with either long(A) or short(B) branches. When there is extensive branching, the polymer can have a dendritic structure in which there are branches protruding from other branches, that is branched branches(C). The presence of branching in a polymer usually has a large effect on many important polymer properties. The most significant changes brought about by branching are the crystallization capability and flow property. Branched polymers do not pack as easily into a crystal lattice as do linear polymers, leading to decrease of crystallinity. It is important to point out that the term branched polymer does not refer to linear polymers containing side groups that are part of the monomer structure. Only those polymers that contain side branches composed of complete monomer units are termed branched polymers.

When polymers are produced in which the polymer molecules are linked to each other at points other than their ends, the polymers are said to be crosslinked. Crosslinking can be made to occur during the polymerization process by the use of appropriate monomers. It can also be brought about after the polymerization by various chemical reactions. The crosslinks between polymer chains can be of different lengths depending on the crosslinking method and the specific conditions employed. One can also vary the number of crosslinks so as to obtain lightly or highly crosslinked polymers. When the number of corsslinks is sufficiently high, a three-dimensional or space network polymer is produced in which all the polymer chains in a sample have been linked together to form one giant molecule. Light crosslinking is used to impart good recovery (elastic) properties to polymers to be used as rubbers. High degrees of crosslinking are used to impart high rigidity and dimensional stability (under conditions of heat and stress) to polymers such as the phenol-formaldehyde and urea-formaldehyde polymers.

3. 2. 4 Classification by Thermal Behavior

Some polymers become soft and fluid upon heating, and then can be converted into any desired shape, which is solidified by subsequent cooling. This cycle can be repeated several times at will. Such polymers, which soften on heating and stiffen on cooling, are termed "thermoplastics". Thermoplastics are composed of polymers with linear or branched chain structures. They account for almost 90 weight percent of the processed polymers. PE, PVC, nylon, PS are examples of thermoplastics polymers.

On the other hand, some polymers undergo chemical change (crosslinking or curing) on heating and convert themselves into an infusible and insoluble mass. They are called "thermosetting" polymersor "thermoset". Thermosets are network polymers that are heavily crosslinked to give a densethree-dimensional network. They are normally rigid, can not melt on heating and decompose if the temperature is not high enough. The typical examples are epoxies, phenolics, unsaturated polyesters, etc. Thermosetting polymers can only be shaped in a single, discontinuous step by compression or injection moulding when the chemical reaction is happening.

Vulcanized rubber or synthetic elastomers are properly crosslinked in order to preserve elasticity, but they are neither meltable nor soluble. In a proper solvent they only swell. In fact, these materials can be considered as thermoplastics before crosslinking and as thermosets after wards.

3. 2. 5 Classification by Polymerization

Polymers were classified into addition polymers and condensation polymers based on the involved synthetic reaction of either addition polymerization or condensation polymerization. Addition polymers are those obtained through addition polymerization. They do not contain functional groups as part of the polymer chain. Such groups may, however, be present in addition polymers as pendant substituent hanging off the polymer chain. According to this classification, the polyurethanes are readily and more correctly classified as condensation polymers. However, with or without functional groups in the polymer chain can not be taken as the only criterion for polymer classification. A polymer is classified as a condensation polymer if its synthesis involves the elimination of small molecules, or it contains functional groups as part of the polymer chain, or its repeating unit lacks certain atoms that are present in the (hypothetical) monomer to which it can be degraded. Thus the structure of condensation polymers has been defined by

$$\left[R^1—Z—R^2 \right]_n \tag{3.9}$$

where R^1 and R^2 are aliphatic or aromatic grouping unit and Z is a functional unit such as —OCO—, —NHCO—, —S—, —OCONH—, —O—, —OCOO—, and—SO_2—.

Some condensation polymers do not contain functional groups such as ester or amide in the polymer chain, such as phenol-formaldehyde polymers produced by the reaction of phenol with formaldehyde. It really is condensation polymer, since water is split out during the polymerization process. Another example is poly (*p*-xylene), which is produced by the oxidative coupling (dehydrogenation) of *p*-xylene:

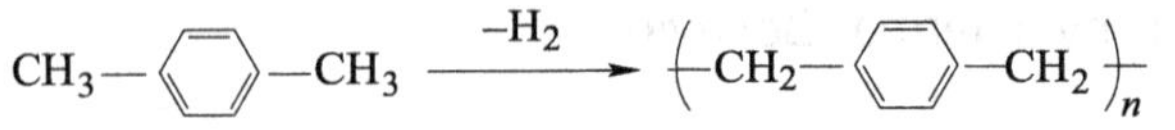
(3.10)

In addition to these classifications mentioned above, there are still other ways to classify polymers, such as by crystallization capability (crystalline polymer and amorphous polymer), or by light penetration (opaque, transparent, translucent, luminescent).

3.2.6 Classification by Elemental Composition of Main Chain

According to the chain chemistry, carbon-chain polymers have a backbone composed entirely of carbon atoms. Most polyolefin belong to this type, e.g. polyethylene, polypropylene, polyvinyl chloride and polystyrene. In contrast, heterochain polymers have other elements in the backbone, such as oxygen in a polyether, —C—O—C—, polyester, O=C—O—C—, polyamide, O=C—NH—. While for another type, the main chain has no carbon atom intirely but only being made of heteroatom, for example polysilicone resin, —Si—O—, polynitrogen sulfide, —S=N—, this refers to elemental organic polymer.

3.3 Chain Structure and Condensed Structure

Polymer structure mainly includes primary structure, secondary structure, and high-level structure. The chain structure often refers to primary and secondary structures. The primary structure mainly describes the composition of unit, connection mode of unit, spatial configuration, optical isomerism, sequence structure and shape of polymer chain. The composition of unit can be illustrated by carbon-chain, heterochain and elemental organic polymer as 3.2.5 part. "head-tail", "head-head", or "tail-tail" always denote three types of connection modes as shown in Eq. (3.11), of which "head-tail" connection mode is the most common one due to the electron effect and steric effect. The spatial configuration is usually described by geometry isomery (*cis*, *trans*). The *cis* configuration arises when substituent groups are on the same side of a carbon-carbon double bond. *Trans* refers to the substituent on opposite sides of the double bond. For example, the natural rubber is composed of *cis*-polyisoprene, while another natural gum is made of *trans*-polyisoprene, i. e. gutta-percha latex as shown in Eq. (3. 12). Tacticity (isotactic, syndiotactic, atactic) is always used to demonstrate optical isomerism while it is not explained in this part due to it has been clarified in detail in 3. 1. 1. The sequence structure always refers to copolymer mode i. e. , alternating, random, graft and block copolymerization as Fig. 3. 6 shown. Similarly, shape of polymer chain includes linear, branched, and crosslinked polymers and the readers can obtain more information from Fig. 3. 8 and here not need to be explained any more.

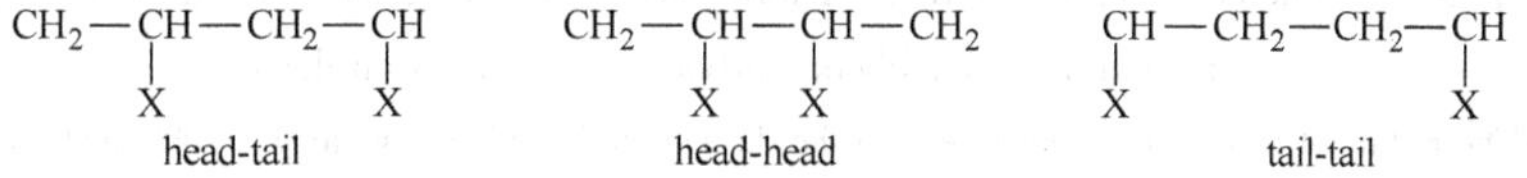

(3. 11)

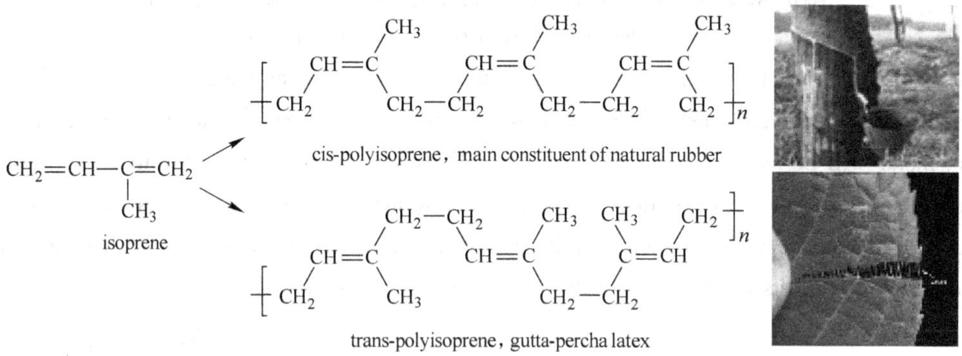

(3. 12)

The secondary structure mainly focus on the size of polymer chain (molecular weight, the molecular weight distribution), conformation of polymer chain and statistic of polymer conformation. When talking about size of polymer chain, Table 3. 1 and Fig. 3. 2 can be given detailed information. The conformations of a molecule are defined as distinct spatial positions adopted by atoms or groups through rotation around single bonds and change in the relative position of a given unit joined by several single bonds. Consider the chain atoms in Fig. 3. 9(a); a third carbon atom may lie at any point on the cone of revolution and still subtend about a 109° with the bond between the other two atoms. A straight chain segment results when successive chain atoms are positioned as in Fig. 3. 9(b). On the other hand, chain bending and twisting are possible when there is a rotation of the chain atoms into other positions, as illustration in Fig. 3. 9(c). Thus, a single chain molecule composed of many chain atoms might assume a shape similar to that represented schematically in Fig. 3. 10, having a multitude of bends, twist, and kinks. Also indicated in this figure is the end-to-end distance of the polymer chain r; this distance is much smaller than the total chain length.

Polymer consist of large numbers of molecular chains, each of which may bend, coil, and kink in the manner of Fig. 3. 10. This leads to extensive intertwining and entanglement of neighboring chain molecules, a situation similar to that of a fishing line that has experienced backlash from a fishing reel. These random coils and molecular entanglements are responsible for a number of important characteristics of polymers, to include the large elastic extensions displayed by the rubber materials.

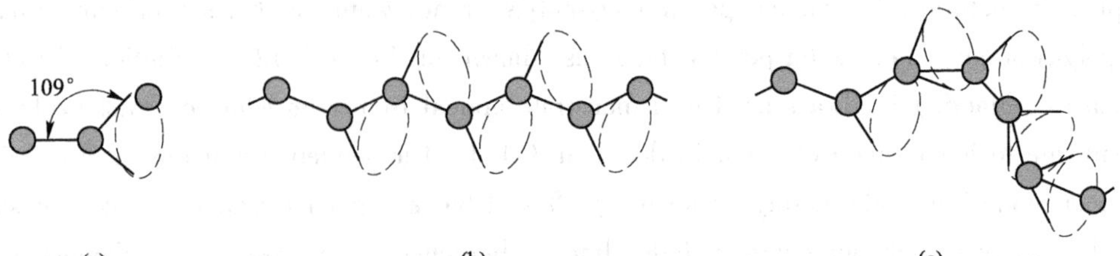

Fig. 3. 9 Schematic representations of how polymer chain shape is influenced by the positioning of backbone carbon atoms (solid circles)

(a) The rightmost atom may lie anywhere on the dashed circle and still subtend a 109° angle with the bond between the other two atoms; (b), (c) Straight and twisted chain segments are generated when the backbone atoms are situated, respectively

The condensed structure of polymer always refers to the crystal and amorphous state. However, since it involves molecules instead of just atoms or ions, as with metals and ceramics, the atomic arrangements will be more complex for polymers. We think of polymer crystallinity as the packing of molecular chains so as to produce an ordered atomic array. Crystal structures may be specified in terms of unit cells, which are often quite complex. For example, Fig. 3. 11 shows the unit cell for polyethylene and its relationship to the molecular chain structure.

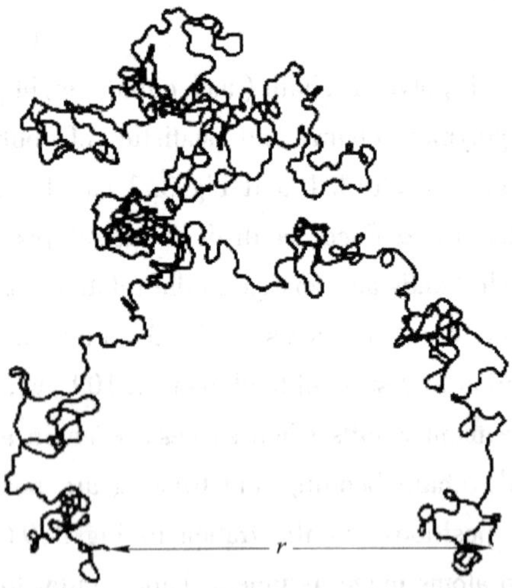

Fig. 3. 10 Schematic representation of a single polymer chain molecule that has numerous random kinks and coils produced by chain bond rotations

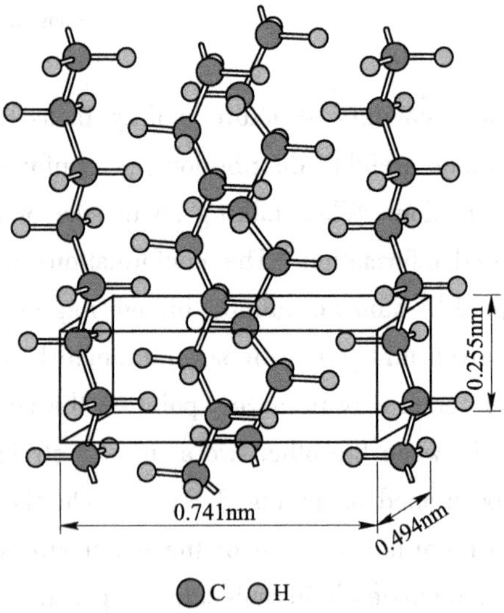

Fig. 3. 11 Arrangement of molecular chains in a unit cell for polyethylene

Molecular substances having small molecules (e. g., water and methane) are normally either totally crystalline (as solids) or totally amorphous (as liquids). As a consequence of their size and often complexity, polymer molecules are often only partially crystalline (or semicrystalline), having crystalline regions dispersed within the remaining amorphous material. Any chain disorder or misalignment will result in an amorphous region, a condition that is fairly common, since twisting, kinking, and coiling of the chains prevent the strict ordering of every segment of every chain. Other structural effects are also influential in determining the extent of crystallinity, as discussed below.

The degree of crystallinity may range from completely amorphous to almost entirely (up to about 95%) crystalline; by way of contrast, metal specimens are almost always entirely crystalline, whereas many ceramics are either totally crystalline or totally noncrystalline. Semicrystalline polymers are, in a sense, analogous to two-phase metal alloys, discussed in subsequent chapters.

The density of a crystalline polymer will be greater than an amorphous one of the same material and molecular weight, since the chains are more closely packed together for the crystalline structure. The degree of crystallinity by weight may be determined from accurate density measurements, according to

$$\text{crystallinity}(\%) = \frac{\rho_c(\rho_s - \rho_a)}{\rho_s(\rho_c - \rho_a)} \times 100 \tag{3.13}$$

where ρ_s is the density of a specimen for which the percent crystallinity is to be determined, ρ_a is the density of the totally amorphous polymer, and ρ_c is the density of the perfectly crystalline polymer. The values of ρ_a and ρ_c must be measured by other experimental means.

The degree of crystallinity of a polymer depends on the rate of cooling during solidification as well as on the chain configuration. During crystallization upon cooling through the melting temperature, the chains, which are highly random and entangled in the viscous liquid, must assume an ordered configuration. For this to occur, sufficient time must be allowed for the chains to move and align themselves.

The molecular chemistry as well as chain configuration also influences the ability of a polymer to crystallize. Crystallization is not favored in polymers that are composed of chemically complex mer structures (e.g., polyisoprene). On the other hand, crystallization is not easily prevented in chemically simple polymers such as polyethylene and polytetrafluoroethylene, even for very rapid cooling rates.

For linear polymers, crystallization is easily accomplished because there are virtually no restrictions to prevent chain alignment. Any side branches interfere with crystallization, such that branched polymers never are highly crystalline; in fact, excessive branching may prevent any crystallization whatsoever. Most network and crosslinked polymers are almost totally amorphous; a few crosslinked polymers are partially crystalline. (With regard to stereoisomers, atactic polymers are difficult to crystallize; however, isotactic and syndiotactic polymers crystallize much more easily because the regularity of the geometry of the side groups facilitates the process of fitting together adjacent chains.) Also, the bulkier or larger the side-bonded groups of atoms, the less tendency there is for crystallization.

3.4 Glass Transition

The glass transition is an important parameter, since many physical properties change when polymer chains gain mobility. This is an important temperature because it marks the point where the polymer becomes more easily deformed, or more ductile. As summarized in Table 3.2, totally non crystalline polymers are very brittle below the glass transition, whereas polymers with intermediate crystallinity are very tough above the glass transition. Most homopolymers have glass transition temperatures that are (0.5~0.75) T_m, even though the crystalline melting point can vary by several hundred degrees between polymer types (see Fig. 3. 12). Most plastics are designed to be used between T_m and T_g, as shown in Fig. 3. 13.

Table 3. 2 Predominant Properties of Crystalline Polymers

Temperature range	Degree of crystallinity		
	Low (5%~10%)	Intermediate (20%~60%)	High (70%~90%)
Above T_g	Rubbery	Leathery, tough	Stiff, hard, brittle
Below T_g	Glassy, brittle	Hornlike, tough	Stiff, hard, brittle

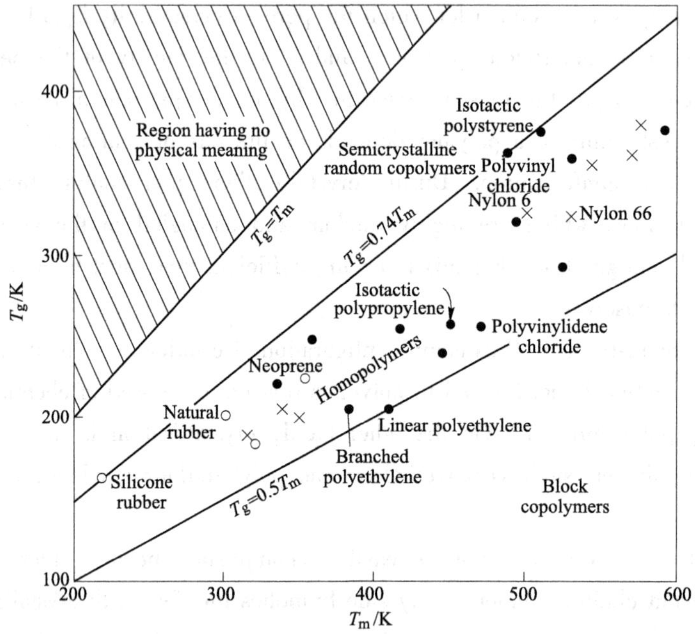

Fig. 3. 12 The glass transition temperature, T_g, as a function of crystalline melting point, T_m, for homopolymers (Filled circles are addition homopolymers, open circles are elastomers, and crosses are condensation homopolymers)

To conclude this section on polymols, we should note that we have used the termpolymer almost exclusively to refer to organic macromolecules. The term plastic refers not only to organic substances of high molecular weight, but also to such substances that at some point in their manufacture have been shaped by flow. Thus, the term plastic is more specific than the term polymer, and this term carries with it an indication of its processing history.

Questions

1. How can you experimentally determine whether the polymerization of an unknown monomer was

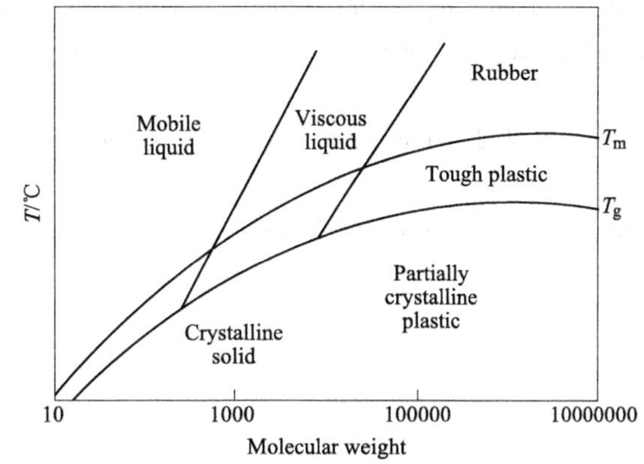

Fig. 3. 13 Approximate relations among molecular weight, T_g, T_m and polymer properties

proceeding by a step or a chain mechanism?

2. Try to finish the following polymerization using them as monomer.
   (a) $CH_2{=}CH{-}COOH$
   (b) $H_2N{-}(CH_2)_5{-}NH_2 + ClCO{-}(CH_2)_5{-}COCl$
   (c) $OH{-}(CH_2)_5{-}COOH$
   (d) 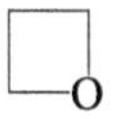
   (e) $CH_2{=}CH{-}F$
   (f) 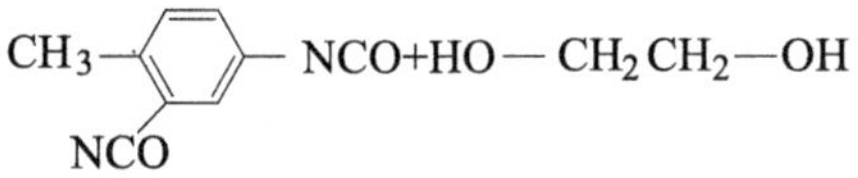

3. Find out one of your interesting polymer materials and share its discovery procedure with your partners.

References

[1] Muelhaupt R. Hermann staudinger and the origin of macromolecular chemistry[J]. Chem. Int. Ed. 2004, 43: 1054~1063.

[2] Bradley D, Fahlman. Materials chemistry[M]. Beijing: Science Press, 2018.

[3] Brian S. Mitchell. An introduction to materials engineering and science for chemical and materials engineers [M]. Hoboken: John Wiley & Sons, 2004.

[4] Heimenz P, Polymer chemistry:The basic concepts[M]. New York:Marcel Dekker, Inc,1984.

[5] William D. Callister, Jr. Fundamentals of materials science and engineering[M]. Hoboken:John Wiley & Sons,2001.

[6] Xiping Lei, Zhixing Su. Synthesis and properties of the copolymers based on aniline and 2-aminobenzyl alcohol [J]. Journal of Polymer Research, 2008, 15:421~426.

[7] Li Jian, Yu Qiang, Wan Tong, et al. Introduction to polymer materials [M]. Beijing: Chemical Industry Press, 2014.

[8] Askeland D R. The science and engineering of materials[M]. 3rd edition. London: Chapman & Hall,1996.

[9] Treloar L R G. The physics of rubber elasticity[M]. 2nd edition. Oxford:Oxford University Press,1958.

[10] Bunn C W. Chemical crystallography[M]. Oxford: Oxford University Press, 1945.

[11] Brian S. Mitchell. An introduction to materials engineering and science[M]. Hoboken: John Wiley & Sons, 2004.

[12] Ralls K M, Courtney T H, Wulff J. Introduction to materials science and engineering[M]. Hoboken: Wiley, 1976.

[13] Billmeyer F W. Textbook of polymer science[M]. 3rd edition. Hoboken: John Wiley & Sons, 1984.

Chapter 4 Preparation Principles in Materials Chemistry

In materials science, the emphasis is on the underlying relationships between the synthesis and processing, structure, and properties of materials. The term "structure" means a description of the arrangement of atoms, as seen at different levels of detail. Materials scientists not only deal with the development of materials, but also with the synthesis and processing of materials and manufacturing processes related to the production of components. The term "synthesis" refers to how materials are made from naturally occurring or man-made chemicals. The term "processing" means how materials are shaped into useful components. The structure of materials has a profound influence on many properties of materials, even if the overall composition does not change! The structure at this microscopic scale is known as micro structure. If we can understand what has changed at a micrometer level, we can begin to discover ways to control the material's properties. Thus, the interrelationship between synthesis, processing, structure and properties is as depicted in the schematic illustration shown in Fig. 4. 1. One of the most fundamental aspects of materials science involves the investigation into the preparation of a material.

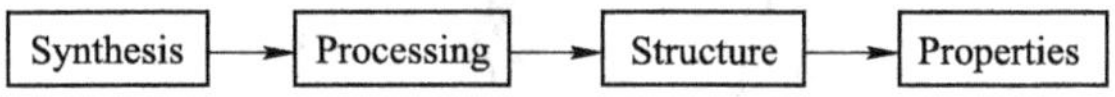

Fig. 4. 1 The interrelationship of four components in the materials chemistry

4. 1 Features of Basic Material

Generally, basic materials have been conveniently grouped into three basic categories: metals, ceramics, and polymers. This scheme is based primarily on chemical makeup and atomic structure, and most materials fall into one distinct grouping or another. In addition, there are the composites, which are engineered combinations of two or more different materials. A brief explanation of these material classifications and representative characteristics is offered next.

4. 1. 1 Metals

Metals in this group are composed of one or more metallic elements (e. g. , iron, aluminum, copper, titanium, gold, and nickel), and often also nonmetallic elements (e. g. , carbon, nitrogen, and oxygen) in relatively small amounts. Atoms in metals and their alloys are arranged in a very orderly manner and are relatively dense in comparison to the ceramics and polymers. With regard to mechanical characteristics, these materials are relatively stiff, yet are ductile (i. e. , capable of large amounts of deformation without fracture) and are resistant to fracture, which accounts for their widespread use in structural applications. Metallic materials have large numbers of non-localized electrons; that is, these electrons are not bound to particular atoms. Many properties of metals are directly attributable to these electrons. For example, metals are extremely good conductors of electricity and heat and are not transparent to visible light; a polished metal surface has a lustrous appearance. In addition, some of the metals (i. e. , Fe, Co, and Ni) have desirable magnetic properties.

In the metals and their alloys (see Fig. 4. 2), metallic bonding is the primary bonding type. A relatively simple model has been proposed that very nearly approximates the bonding scheme. Metallic materials have one, two, or at most three valence electrons. With this model, these valence electrons are not bound to any particular atom in the solid and are more or less free to drift throughout the entire metal. They may be thought of as belonging to the metal as a whole, or forming a "sea of electrons" or an "electron cloud". The remaining non-valence electrons and atomic nuclei form what are called ion cores, which possess a net positive charge equal in magnitude to the total valence electron charge per atom. Fig. 4. 3 is a schematic illustration of metallic bonding. The free electrons shield the positively charged ion cores from the mutually repulsive electrostatic forces that they would other-wise exert upon one another; consequently the

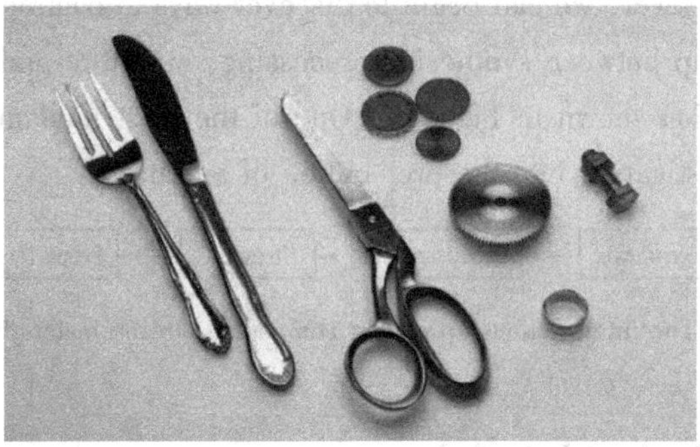

Fig. 4. 2 Familiar objects that are made of metals and metal alloys

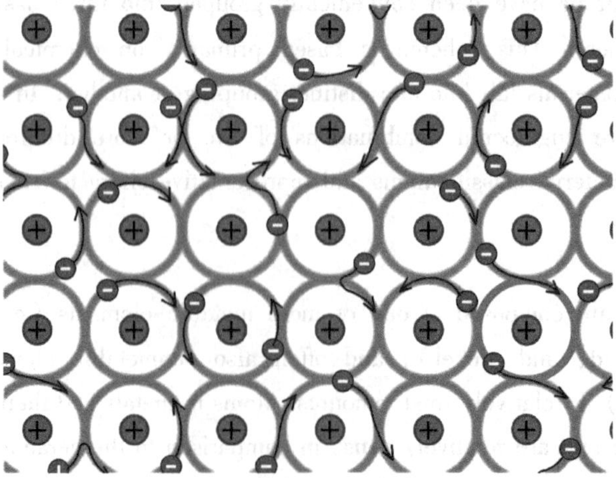

Fig. 4. 3 Schematic illustration of metallic bonding

metallic bond is non-directional in character. In addition, these free electrons act as a "glue" to hold the ion cores together. Bonding may be weak or strong; energies range from 68kJ/mol (0.7eV/atom) for mercury to 849kJ/mol (8.8eV/atom) for tungsten. Their respective melting temperatures are −39℃ and 3410℃ (−38℉ and 6170℉). Metallic bonding is found in the periodic table for Group ⅠA and ⅡA elements and, in fact, for all elemental metals. Some general behaviors of the various material types (i.e., metals, ceramics, polymers) may be explained by bonding type. For example, metals are good conductors of both electricity and heat, as a consequence of their free electrons. By way of contrast, ionically and covalently bonded materials are typically electrical and thermal insulators because of the absence of large numbers of free electrons. Furthermore, most metals and their alloys fail in a ductile manner; that is, fracture occurs after the materials have experienced significant degrees of permanent deformation. This behavior is explained interms of deformation mechanism, which is implicitly related to the characteristics of the metallic bond.

The atomic bonding in this group of materials is metallic and thus nondirectional in nature. Consequently, there are minimal restrictions as to the number and position of nearestneighbor atoms; this leads to relatively large numbers of nearest neighbors and dense atomic packings for most metallic crystal structures. Also, for metals, when we use the hard-sphere model for the crystal structure, each sphere represents an ion core. Table 4.1 presents the atomic radii for a number of metals. Three relatively simple crystal structures are found for most of the common metals: face-centered cubic, body-centered cubic, and hexagonal close-packed.

Table 4.1 Atomic radii and crystal structures for 16 metals

Metal	Crystal structure[a]	Atomic radius[b]/nm	Metal	Crystal structure[a]	Atomic radius[b]/nm
Aluminun	FCC	0.1431	Molybdenum	BCC	0.1363
Cadmium	HCP	0.1490	Nickel	FCC	0.1246
Chromium	BCC	0.1249	Platinum	FCC	0.1387
Cobalt	HCP	0.1253	Silver	FCC	0.1445
Copper	FCC	0.1278	Tantalum	BCC	0.1430
Gold	FCC	0.1442	Titanium(α)	HCP	0.1445
Iron(α)	BCC	0.1241	Tungsten	BCC	0.1371
Lead	FCC	0.1750	Zinc	HCP	0.1332

[a]FCC = face-centered cubic; HCP = hexagonal close-packed; BCC = body-centered cubic.

[b]A nanometer (nm) equals 10^{-9} m; to convert from nanometers to angstrom units (Å), multiply the nanometer value by 10.

4.1.2 Ceramics

Ceramics are compounds between metallic and nonmetallic elements (see Fig. 4.4); they are most frequently oxides, nitrides, and carbides. For example, common ceramic materials include aluminum oxide (or alumina, Al_2O_3), silicon dioxide (or silica, SiO_2), silicon carbide (SiC), silicon nitride (Si_3N_4), and, in addition, what some refer to as the traditional ceramics—those composed of clay minerals (i.e., porcelain), as well as cement and glass. With regard to mechanical behavior, ceramic materials are relatively stiff and strong-stiffnesses and strengths are comparable to those of the metals. In addition, they are typically very hard. Historically, ceramics have exhibited extreme brittleness (lack of ductility) and are highly susceptible to fracture. However, newer ceramics are being engineered to have improved resistance to fracture; these materials are used for cookware, cutlery, and even automobile engine parts. Furthermore, ceramic materials are typically insulative to the passage of heat and electricity (i.e., have low electrical conductivities;) and are more resistant to high temperatures and harsh environments than are metals and polymers. With regard to optical characteristics, ceramics may be transparent, translucent, or opaque, and some of the oxide ceramics (e.g., Fe_3O_4) exhibit magnetic behavior.

Fig. 4. 4 Common objects that are made of ceramic materials

Ceramics are composed of at least two elements and often more, their crystal structures are generally more complex than those of metals. The atomic bonding in these materials ranges from purely ionic to totally covalent; many ceramics exhibit a combination of these two bonding types, the degree of ionic character being dependent on the electronegativities of the atoms. These values were determined using Eq. (4.1) and the electronegativities.

$$IC(\%) = \{1 - \exp[-0.25(X_A - X_B)^2]\} \times 100 \quad (4.1)$$

where X_A and X_B are the electronegativities for the respective elements. For those ceramic materials for which the atomic bonding is predominantly ionic, the crystal structures may be thought of as being composed of electrically charged ions instead of atoms. The metallic ions, or cations, are positively charged because they have given up their valence electrons to the nonmetallic ions, or anions, which are negatively charged. Two characteristics of the component ions in crystalline ceramic materials influence the crystal structure: the magnitude of the electrical charge on each of the component ions, and the relative sizes of the cations and anions. With regard to the first characteristic, the crystal must be electrically neutral; that is, all the cation positive charges must be balanced by an equal number of anion negative charges. The chemical formula of a compound indicates the ratio of cations to anions, or the composition that achieves this charge balance. For example, in calcium fluoride, each calcium ion has a +2 charge (Ca^{2+}), and associated with each fluorine ion is a single negative charge (F^-). Thus, there must be twice as many F^- as Ca^{2+} ions, which is reflected in the chemical formula CaF_2.

In covalent bonding, stable electron configurations are assumed by the sharing of electrons between adjacent atoms. Two atoms that are covalently bonded will each contribute at least one electron to the bond, and the shared electrons may be considered to belong to both atoms. Covalent bonding is schematically illustrated in Fig. 4. 5 for a molecule of methane (CH_4). The carbon atom has four valence electrons, whereas each of the four hydrogen atoms has a single valence electron. Each hydrogen atom can acquire a helium electron configuration (two 1s valence electrons) when the carbon atom shares with it one electron. The carbon now has four additional shared electrons, one from each hydrogen, for a total of eight valence electrons, and the electron structure of neon. The covalent bond is directional; that is, it is between specific atoms and may exist only in the direction between one atom and another that participates in the electron sharing.

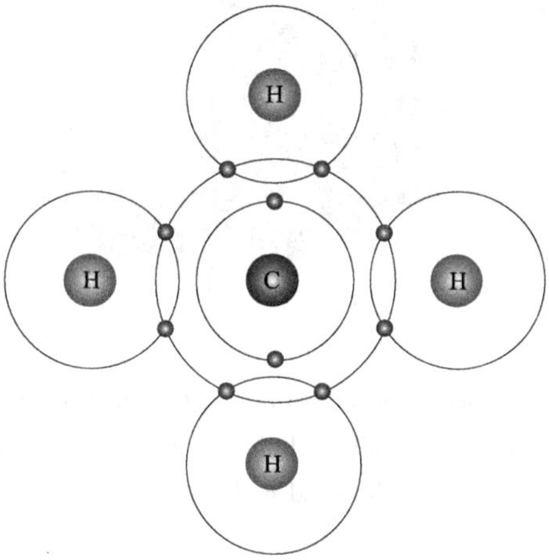

Fig. 4. 5 Schematic representation of covalent bonding in a molecule of methane (CH_4)

Many nonmetallic elemental molecules (H_2, Cl_2, F_2, etc.), as well as molecules containing dissimilar atoms, such as CH_4, H_2O, HNO_3, and HF, are covalently bonded. Furthermore, this type of bonding is found in elemental solids such as diamond (carbon), silicon, and germanium and other solid compounds composed of elements that are located on the right-hand side of the periodic table, such as gallium arsenide (GaAs), indium antimonide (InSb), and silicon carbide (SiC).

4. 1. 3 Polymeric Materials

Polymers include the familiar plastic and rubber materials (see Fig. 4. 6). Many of them are organic compounds that are chemically based on carbon, hydrogen, and other nonmetallic elements (i. e. , O, N, and Si). Furthermore, they have very large molecular structures, often chain-like in nature, that often have a backbone of carbon atoms. Some common and familiar polymers are polyethylene (PE), nylon, poly(vinyl chloride) (PVC), polycarbonate (PC), polystyrene (PS), and silicone rubber. These materials typically have low densities, whereas their mechanical characteristics are generally dissimilar to those of the metallic and ceramic materials—they are not as stiff or strong as these other material types. However, based on their low densities, many times their stiffnesses and strengths on a per-mass basis are comparable to those of the metals and ceramics. In addition, many of the polymers are extremely ductile and pliable (i. e., plastic), which means they are easily formed into complex shapes. In general, they are relatively inert chemically and unreactive in numerous environments. One major drawback to the polymers is their tendency to soften and/or decompose at modest temperatures, which, in some instances, limits their use. Furthermore, they have low electrical conductivities and are nonmagnetic.

Fig. 4. 6 Several common objects that are made of polymeric materials: plastic tableware (spoon, fork, and knife), billiard balls, a bicycle helmet, two dice, a lawnmower wheel (plastic hub and rubber tire), and a plastic milk carton

The physical characteristics of a polymer depend not only on its molecular weight and shape, but also on differences in the structure of the molecular chains. Modern polymer synthesis techniques permit considerable control over various structural possibilities. This section discusses several molecular structures including linear, branched, crosslinked, and network, in addition to various isomeric configurations.

Linear polymers are those in which the repeat units are joined together end to end in single chains. These long chains are flexible and may be thought of as a mass of "spaghetti", as represented schematically in Fig. 4. 7(a), where each circle represents a repeat unit. For linear polymers, there may be extensive van der Waals and hydrogen bonding between the chains. Some of the common polymers that form with linear structures are polyethylene, poly(vinyl chloride), polystyrene, poly (methyl methacrylate), nylon, and the fluorocarbons. Polymers may be synthesized in which side-branch chains are connected to the main ones, as indicated schematically in Fig. 4. 7 (b); these are fittingly called branched polymers. The branches, considered to be part of the main-chain molecule, and may result from side reactions that occur during the synthesis of the polymer. The chain packing efficiency is reduced with the formation of side branches, which results in a lowering of the polymer density. Polymers that form linear structures may also be branched. For example, high-density polyethylene (HDPE) is primarily a linear polymer, whereas low-density polyethylene (LDPE) contains short-chain branches. In crosslinked polymers, adjacent linear chains are joined one to another at various positions by covalent bonds, as represented in Fig. 4.7(c). The process of crosslinking is achieved either during synthesis or by a nonreversible chemical reaction. Often, this crosslinking is accomplished by additive atoms or molecules that are covalently bonded to the chains. Many of the rubber elastic materials are crosslinked; in rubbers, this is called vulcanization. Multifunctional monomers forming three or more active covalent bonds make three-dimensional networks (Fig. 4.7(d)) and are termed network polymers. Actually, a polymer that is highly crosslinked may also be classified as a network polymer. These materials have distinctive mechanical and thermal properties; the epoxies, polyurethanes, and phenol-formaldehyde belong to this group. Polymers are not usually of only one distinctive structural type. For example, a predominantly linear polymer may have limited branching and crosslinking.

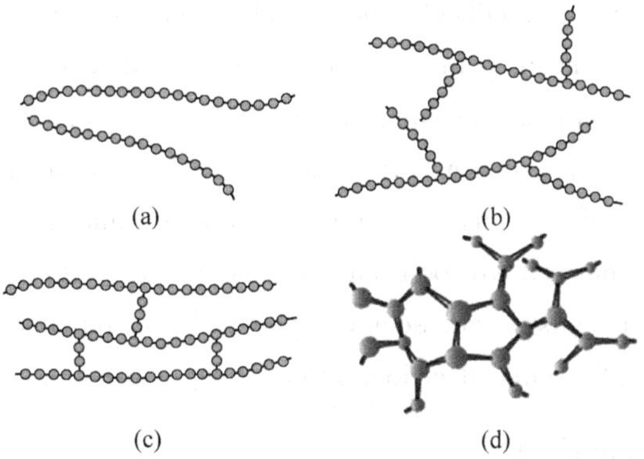

Fig. 4.7 Schematic representations of polymers (Circles designate individual repeat units)
(a) Linear; (b) Branched; (c) Crosslinked; (d) Network (three-dimensional) molecular structures

4.1.4 Materials Design and Selection

When a material is designed for a given application, a number of factors must be considered. The material must possess the desired physical and mechanical properties. It must be capable of being processed or manufactured into the desired shape, and must provide an economical solution to the design problem. Satisfying these requirements in a manner that protects the environment—perhaps by encouraging recycling of the materials—is also essential. In meeting these design requirements, the engineer may have to make a few of tradeoffs in order to produce a serviceable, yet marketable, product.

As an example, material cost is normally calculated on a cost-per-pound basis. We must consider the density of the material, or its weight-per-unit volume, in our design and selection. Aluminum may cost more per pound than steel, but it is only one-third the weight of steel. Although parts made from aluminum may have to be thicker, the aluminum part may be less expensive than the one made from steel because of the weight difference.

In some instances, particularly in aerospace applications, the weight issue is critical, since additional vehicle weight increases fuel consumption and reduces range. By using materials that are lightweight but very strong, aerospace or automobile vehicles can be designed to improve fuel efficiency. Many advanced aerospace vehicles use composite materials instead of aluminum alloys. These composites, such as carbon-epoxy, are more expensive than the traditional aluminum alloys; however, the fuel savings yielded by the higher strength-to-weight ratio of the composite may offset the higher initial cost of the aircraft. The body of one of the latest Boeing aircrafts known as the Dreamliner is made almost entirely from carbon-carbon composite materials. There are literally thousands of applications in which similar considerations apply. Usually, the selection of materials involves trade-offs between many properties.

By this point of our discussion, we hope that you can appreciate that the properties of materials depend not only on composition, but also on how the materials are made (synthesis and processing) and, most importantly, their internal structure. This is why it is not a good idea for an engineer to simply refer to a handbook and select a material for a given application. The handbooks may be a good starting point. A good engineer will consider: the effects of how the material is made, what exactly is the composition of the candidate material for the application being considered, any processing that may have to be done for shaping the material or fabricating a component, the structure of the material after processing into a component or device, the environment in which the material will be used, and the cost-to-performance ratio. The knowledge of principles of materials chemistry will empower you with the fundamental concepts. These will allow you to make technically sound decisions in designing with engineered materials.

4.2 Solid State Reaction

4.2.1 Definition and Features

Reactions between or within solid reactants to yield a solid product are prototypical of solvent free reactions. From literature reports the concept of chemical reactivity between solids (and very often within solids) is very difficult to define. Reactions which occurred in the melt were not considered as genuine solid-state reactions. The same applied to reactions which occurred with crystal degradation prior to the reaction. For the purpose of this thesis, solid state reactions shall refer to all solvent-less processes leading from a solid reactant to a solid product. Gas solid reactions will also be regarded as solid-state reactions as they can eventually produce a solid product. Solid state reactions are performed in the absence of solvents by either grinding or melting the starting materials together or simply applying heat to a mixture of starting materials. This type of reaction is usually performed in order to obtain polycrystalline inorganic solids but may also be used in organic synthesis. Solid state reactions offer reduced costs, decreased amounts of chemical waste and, sometimes, an increase in yield. In all type solid state reaction, at least one of the reactants must diffuse towards the other reactants, which is shown in the Fig. 4.8.

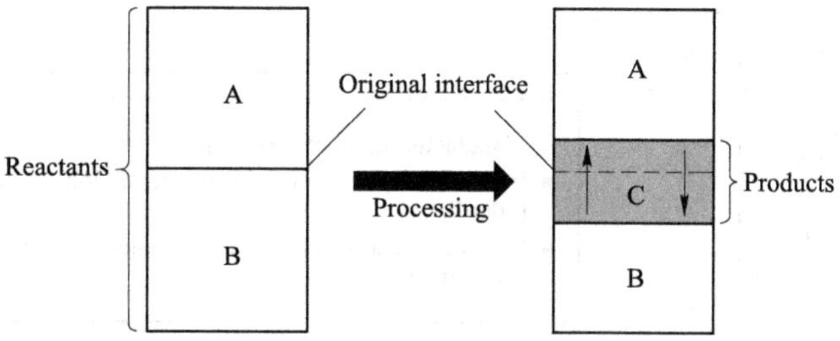

Fig. 4. 8 Schematic illustration of solid state reactions

The difference between solid state reactions and solution ones is that solid state reactions occur within the rigid constraining environment of the crystal lattice (reaction cavity concept). Solid state reactions therefore provide an extreme case for evaluating the effect of intermolecular forces on a reaction and their influence on reaction mechanism and direction. The confined environment of the reactant crystal lattice can control the kinetic features of a reaction and hence the nature of products. Such reactions are said to be controlled and illustrate the topochemical principle proposed by Cohen and Schmidt in which reactions in the solid state take place with minimum requirements of energy and atomic or molecular motion requirements. Topochemical control of a reaction to produce a product is analogous to the kinetic control of a reaction in solution. The product is not necessarily the thermodynamically most stable product available to the system, but is rather the one dictated by the reaction pathway available in the encapsulating surroundings of the solid. Thus product selectivity in the solid state can be different from reactions in solution or gaseous media since the range of intermediates or transition state structures accessible in a confined environment are more limited. Regardless of whether or not topochemical control is evident in a reaction, there is often correlation at the molecular level, between the lattice of the product and that of the reactant. When this occurs the reaction is then described as being topotactic. In other instances the growth of the product phase proceeds in a less orderly manner, producing a product that is amorphous to X-rays. This situation seems to arise when the product lattice fractures and disrupts the host lattice such that long range coherence within the crystal lattice is lost.

Generally, solid state reactions may have the following advantages over solution phase reactions:

(1) The atom economic nature of the solid-state reactions and the limited formation of side products.

(2) No solvents are needed in the reaction and hence no waste disposal issues associated with the solvent need be considered. This is of considerable importance in the rapidly emerging field of Green Chemistry which has resulted in major changes in the way synthetic chemists develop processes and products. Also, products do not require extensive purification to remove traces of solvent and impurities, which is shown in the Fig. 4. 9.

(3) The constrained environment in a "reaction cavity" can lead to novel chemical reactions, with product selectivity different from that expected in the liquid phase. It is expected that the crystalline solid state may affect the product selectivity of chemical reactions, not only by rate

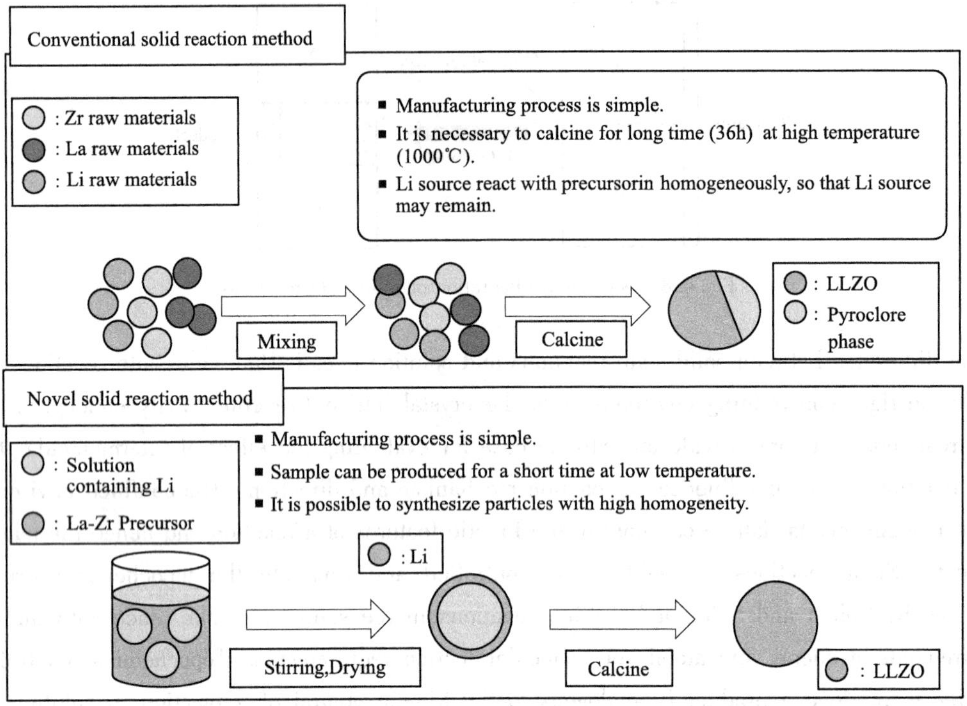

Fig. 4. 9 Conceptual representation of novel approach to solid state synthesis approach in contrast to conventional solid reaction method

accelerations but also by slowing down rates of competing processes. Manipulating rates and selectivity by changing activation energies in the liquid phase can be thought of as a low-resolution approach to controlling chemical reactivity. It is therefore possible that solid state reactions may occur with both control and yields not achievable by any other media, indeed quantitative yields have been observed with solid state reactions. These reactions are often faster, taking just a few minutes rather than hours to complete because the reactants are in intimate contact with each other.

(4) Enantioselective reactions are feasible in a chiral cavity. In such reactions the asymmetry is only determined by its physical origin without any chemical reagent—the chiral source being the chiral crystal packing. Kinetic trapping of highly reactive intermediates.

(5) Crystal engineering using organometallic complexes is possible in the solid state. Cascade reactions can be carried out quite easily and with high yields. The potential of solid-state chemistry for mechanistic studies, solvent free synthesis and materials applications is now fully recognized.

4. 2. 2 Solid-solid Interface

In this part, we examine the solid-solid interface and review some solid-solid synthetic techniques, neither solid will remain chemically identifiable after the reaction. Moreover, the slow time scale on which atoms in the solid phase move will require high temperatures and small reaction distances for reasonable reaction times. Before discussing a few specific synthetic techniques, we examine the generic solid-solid interface. For simplicity, consider a solid-solid interface composed of two crystalline materials, as shown in Fig. 4.10. The most obvious feature of the solid-solid interface is that both phases are tightly packed, and one phase moving into the other, vacancies or interstices are necessary. Moving atoms in one phase out of the way to accommodate atoms from the other phase will take a great deal of energy. The diffusion coefficient for atoms in this scenario is low. Diffusion coefficient, D, for atoms in a solid are many orders of magnitude less than what is observed in liquids and gases. Typical values are of D are: gases, $10^{-1}cm^2/s$; liquids, $10^{-5}cm^2/s$; and solids, $10 \sim 20cm^2/s$. The units on the diffusion coefficient, cm^2/s, are such that Dt, where t is time in seconds, gives the square of the average distance traveled by a particular atom. Hence, in 1 minute, a molecule in a liquid with $D = 10^{-5}cm^2/s$ will diffuse, on average, 0.02cm. In a solid, however, on the same time scale, an atom will diffuse only $7.7 \times 10^{-10}cm$, or about 0.01nm. In Fig. 4.10, the distance between the atoms û is typically 0.2nm. Because there is an energy barrier to diffusion, the diffusion û coefficient has a temperature dependence similar to the Arrhenius behavior observed in chemical reaction rates: namely,

$$D(T) = D(0)e^{-E_a/(RT)} \tag{4.2}$$

where E_a is the activation energy for diffusion. This equation, and the one relating displacement to D, form the basis for all solid-solid synthetic strategies.

Fig. 4.10 Interface of two different crystals prior to a chemical reaction occurring

Because diffusion has an activation energy barrier, higher temperatures result in higher diffusion coefficients. Furthermore, smaller particle sizes mean smaller Δx required for a complete reaction. Therefore, the traditional approach to solid-solid reactions are to use fine powders and high reaction temperatures. This direct combination of two or more powdered solids at high temperature, referred to as the ceramic method of synthesis.

It is well known that the deposition in molecular beam epitaxy is kept to small thicknesses, on the order of a monolayer, so that a very small diffusion coefficient, D, does not prevent a reaction from occurring on a reasonable timescale. In the ceramic method, we are again presented with the interdependence of the diffusion coefficient, reaction temperature, and displacement distance, Eqs. (4.2) and (4.3).

$$\Delta x^2 = 2Dt \tag{4.3}$$

4.3 Liquid State Reaction

Now, we discuss preparative routes for inorganic materials in three basic types of systems involving the presence of a distinct solid-liquid interface: those in which the liquid and solid phases are of the same chemical identity (solidification and vitrification processes), those in which the liquid and solid phases are not of the same chemical identity (crystallization, precipitation), and the special case in which the liquid phase is a pure ionic liquid or molten salt. Ionic liquids can serve as the solvent as well as a templating agent, and the liquid components may or may not become incorporated into the final solid product. We also discuss two areas where the distinct solid-liquid interface becomes somewhat blurred: namely, sol-gel and solvothermal processes.

In many respects, crystallization from the molten state is analogous to that from a solvent phase. For example, in both processes slow cooling tends to result in larger crystals, while faster cooling typically gives smaller crystals. Another important similarity involves the behavior of impurities. Impurity atoms usually do not "fit" into the crystal lattice of the solute crystallizing from a solvent or melt. In a polycrystalline metal, impurities tend to segregate at grain boundaries during solidification because of their mismatch with the lattice of the metal atoms. This behavior allows crystallization from a solvent to be used as a purification technique, and it is the basis of zone-melting and zone-refining techniques used in the production of very high purity monocrystalline materials. However, there are important differences between the solid-melt and solid-solvent interface as well.

The cohesive forces between the particles in a liquid are intermediate in strength compared to those in a solid or gas. The particles of a liquid have more freedom of movement, and hence more kinetic energy and internal energy than those of a solid, but less than those of a gas. As a direct consequence of this greater energy, there are two major structural differences between a solid and a liquid. First, liquids exhibit variation in short-range order. For example, a liquid network may be described as being comprised of various types of polyhedral "holes", with the constituent atoms or molecules at the vertices. Examples of these polyhedral include tetrahedra, octahedra, trigonal prisms, and Archimedean antiprisms. The second key structural feature differentiating liquids from solids is the absence of long-range order or, more precisely, translational symmetry. In both conventional (i. e., non-glass-forming) solidification and crystallization processes, liquid-phase solute atoms or molecules become rearranged into a more favorable, lower-energy, crystalline phase with long-range order. In fact, that portion of the liquid phase itself in the vicinity of the solid-liquid interface exhibits crystal-induced ordering in the first three to four atomic layers away from the interface. Much less is known about the precise structures of ionic liquids and molten salts. However, some experimental evidence exists suggesting that there are strong similarities between the crystalline and liquid arrangements.

4.3.1 Ostwald Ripening

Although there are analogies between solidification and crystallization, there are also some important structural differences between the solid-melt interface (also called the solidification front) and the solid-solvent interface due to the differing concentration, viscosity, and temperature of the respective liquid phase. These may be summarized as follows:

(1) A conventional melt is a nominally pure phase for which the thermodynamic activity is constant (equal to unity), whereas a solute of low concentration in a solvent has an activity less than the corresponding Gibbs free energy of the pure phase.

(2) The melt viscosity determines the likelihood that a glass will form with rapid cooling. In highly viscous melts, the atomic mobility is reduced substantially, suppressing the homogeneous nucleation rate and crystallizability, which are dependent on the ease with which atomic rearrangement can occur.

(3) The effect of temperature on liquid structure was pioneered by John Desmond Bernal, who suggested that the effect of heating a liquid could be described as a volume expansion of the interstices (holes) existing between the constituent atoms while maintaining the same minimal nearest-neighbor distance between coordinated atoms. For example, a heated liquid contains a greater percentage of larger polyhedral holes (e. g., trigonal prisms, Archimedean antiprisms) than smaller polyhedral holes (e. g., tetrahedra, octahedra).

In 1897, Friedrich Wilhelm Ostwald published his now famous study of crystallization processes, which led to the *Ostwald rule of stages or Ostwald step rule*. Ostwald noticed that the course of transformation of unstable (or metastable) states into stable states normally occurs in stages, where the first products formed are usually not the most thermodynamically stable products but rather, intermediate products having free energies closest to the initial states. Take, for example, the crystallization of citric acid. Crystallization above 34℃ leads to a thermodynamically stable anhydrous crystal, whereas crystallization at roomtemperature leads to a monohydrate polymorph. In detailed crystallization experiments, a probing of the metastable border between the two polymorphs followed the solubility curve of anhydrous citric acid.

In subsequent experiments, using other crystal systems, such as ferrous sulfate and sodium hydrogen phosphate, it was similarly observed that the first crystallization product to form was the one most closely resembling the structure of the solvent. For the case of citric acid, this is the monohydrate, which more closely resembles the aqueous structure. As the temperature of the solution is increased, the structure of the solvent, as well as the solubility of the crystal, changes, resulting in a more thermodynamically stable anhydrous product. This conversion between the kinetic and thermodynamic product occurs at a critical transition temperature, below which the structure of the solution favors the formation of the hydrated product. As the transition temperature is surpassed, the anhydrous product becomes favored. Another important issue relating to metastability is the contribution of the interfacial free-energy to the total free energy of a system. The interfacial free-energy contribution is negligible for crystals larger than 1μm. By a process occurring near equilibrium conditions, known as Ostwald ripening, larger clusters grow at the expense of dissolving smaller clusters. Hence, although many small crystals may form in a system initially, most of them slowly disappear, except for a few that grow larger by consuming the others. The phenomenon was first described for small molecules by Ostwald in 1896. However, it has a wide range of applicability, from the growth of single crystals in solution under ambient conditions to the coarsening of polycrystalline geochemical samples in a subcritical aqueous phase (water temperature >374℃, pressure>2.239×10^4kPa) and of polycrystalline solidification products. In polycrystals, the larger crystallites, which usually possess a lower surface free energy, destabilize smaller coexisting crystallites of the same phase. This results in the growth of progressively fewer, larger crystallites, as well as a smaller crystallite size distribution. A fully satisfactory physical basis for Ostwald ripening also remains elusive. In addition, although only a few violations have been cited, there is still no undisputed theoretical basis for the Ostwald step rule. If we examine the competitive growth of multiple-size crystallites in solutions, we find the growth occurring along concentration gradients. The concentration gradients around the particles are explained with the Gibbs-Thomson equation:

$$C_r = C_\infty \exp\left(\frac{2\gamma\Omega}{rRT}\right) \tag{4.4}$$

where C_r is the solubility at the surface of a spherical particle with radius r, C_∞ is the solute concentration (mole fraction) at a plane interface at equilibrium with a particle of infinite radius, γ is the specific interfacial energy of the solution-precipitate particle boundary, Ω is the mean atomic volume of a particle, R is the gas constant [8.314×10^3J/(K · kmol)], and T is the absolute temperature. Eq. (4.4) is actually only one of several possible forms of the Gibbs-Thomson equation [named after Josiah Gibbs and three Thomsons: James (1822 ~ 1892), his brother Williams (1824 ~ 1907), and Joseph John (1856 ~ 1940)]; it is also known as the Ostwald-Freundlich equation after Ostwald and Herbert Freundlich (1880 ~ 1941), who had corrected Ostwald's original expression. At equilibrium, $2/r \sim dA/dV$ (A is the particle surface area and V is the particle volume), which implies that small particles are more soluble and hence inherently less stable than larger particles. This observation had actually been made as early as 1813 by William Hyde Wollaston and again in 1885 by Pierre Curie.

4.3.2 Crystal Nucleation and Growth

For several decades, Ostwald's work was the foundation of crystallization theory. In 1950, an American chemist, Victor Kuhn LaMer (1895 ~ 1966), advanced our understanding of crystallization of colloidal sols. LaMer assumed that crystallization was a diffusion-driven process. LaMer's model is described in three steps, presented in Fig. 4.11. In the initial steps of the reaction, the concentration of the molecular materials builds until a critical concentration is reached. This critical concentration, called *supersaturation*, can be written as

$$\Delta\mu = k_B T \ln \frac{C}{C_s} \tag{4.5}$$

where k_B is the Boltzmann constant and T is temperature. Supersaturation is defined via this relationship: namely, as the relationship where

$$\frac{\Delta\mu}{k_B T} = \ln \frac{C}{C_s} \tag{4.6}$$

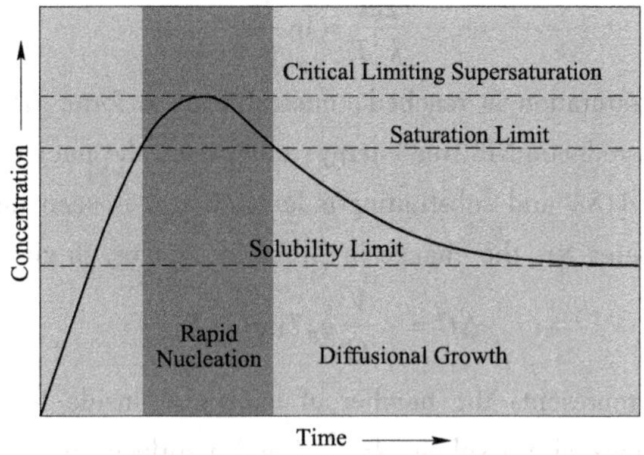

Fig. 4.11 Nucleation process

(As time proceeds, the concentration increases until eventually reaching the supersaturation limit. At this point, nucleation occurs, relieving the high concentration stress.)

At this point, the molecular materials begin to collide and form molecular clusters that are in equilibrium with isolated molecular material. It is the second step that the clusters begin to have enough stability for nucleation to occur, thereby creating the first stable nuclei. At this point, the concentration of free molecular material decreases partially, relieving the supersaturation. The nucleation can occur in two fashions: homogeneous or heterogeneous nucleation. Once the concentration drops below the supersaturation concentration, nucleation can no longer occur and the stable nuclei grow following diffusion kinetics and Ostwald ripening. As the crystal continues to grow, the concentration of free molecular material continues to decrease until eventually, the solubility limit will be reached and the crystal will reach equilibrium with dissolution of the crystal. More recent modeling work has cast doubt on LaMer's simple model for growth of crystals. The primary reason for the LaMer model failure is the principal assumption that the growth is a diffusion-controlled process relying only on unipolar reactions. In more recent work, reactions are found to consist of multiple steps where the concentration of molecular material is being generated, continually resulting in a significantly higher concentration around the supersaturation limit. This leads to multiple nucleation events and a wide size of crystals being formed.

The reversible work, W, to homogeneously nucleate a cluster of a new phase in (unstable) equilibrium with the parent phase is the change in the Gibbs free energy, ΔG:

$$W = \Delta G = - n \times \Delta\mu + S\gamma \tag{4.7}$$

In this equation, n is the number of molecules in the cluster, $\Delta\mu$ is the difference of the bulk chemical potentials of the two phases, S is the area of the cluster, and γ is the interfacial free energy. Eq. (4.7) applies to both homogeneous nucleation in solidification and crystallization from a saturated solution. However, if we are talking about homogeneous nucleation of a solute from a saturated solution, the difference between the chemical potential of a solute in the supersaturated and saturated solutions is given by Eq. (4.5). Supersaturation is defined via are arrangement of Eq. (4.5) [i. e., Eq. (4.7)], which is given again here:

$$\frac{\Delta\mu}{k_B T} = \ln \frac{C}{C_s} \tag{4.8}$$

When sufficient supersaturation is reached, nuclei begin to form. The formation of a nucleus must be accompanied by reduction in free energy; otherwise, the nucleus just returns to solution. Using Eqs. (4.5) and (4.8) and substituting β for C/C_s, it is seen that in order to nucleate a cluster of volume V and area S, the change in Gibbs free energy is given by

$$\Delta G = -\frac{V}{\Omega} k_B T \ln\beta + S\gamma \tag{4.9}$$

The coefficient V/Ω represents the number of molecules inside the nucleus. If we initially consider a spherical nucleus with a volume $4\pi r^3/3$ and a surface area $4\pi r^2$, Eq. (4.9) becomes

$$\Delta G = -\frac{4\pi r^3}{\Omega} k_B T \ln\beta + S\gamma \tag{4.10}$$

The cost associated with a change in energy becomes smaller as the particle surface/volume ratio decreases. Since the volume and surface terms are in competition, there is a certain radius beyond which growth is favored, given by

$$r^* = \frac{2\Omega\gamma}{k_B T \ln\beta} \tag{4.11}$$

This radius is called the critical radius, for which the critical activation energy for nucleation, ΔG^*, is

$$\Delta G^* = \frac{16\pi\Omega^2\gamma^3}{3(k_B T \ln\beta)^2} \tag{4.12}$$

By utilizing Eq. (4.11), Eq. (4.12) is also seen to be

$$\Delta G^* = \frac{4\pi r^{*2}\gamma}{3} \tag{4.13}$$

which shows that the activation energy required to nucleate a cluster, stable at supersaturation β, is one-third the energy required to create its surface. If one solute particle is withdrawn from the nucleus, the cluster dissolves. If one is added, the cluster continues to grow spontaneously. As the temperature is increased, or as the concentration is increased past supersaturation, the activation energy required decreases. Thus, we would expect a faster nucleation rate at higher concentrations or higher temperatures. It is worth noting that the $16\pi/3$ in Eq. (4.12) is the shape factor specific to a sphere. For cubic nuclei, $V = 8r^3$ and $S = 24r^2$ in Eq. (4.10). We can thus see that it is easiest to think of nucleation in terms of two competing processes, the coalescing and redissolution of clusters. Solute particles coalesce spontaneously to form small clusters (e.g., dimers, trimers), but as long as the cluster size is below the critical size, it is not thermodynamically favored, so it redissolves. However, stable numbers of solute particles tend to stay associated. These stable numbers are often called the *magic numbers*. Moreover, they represent the most stable geometric configurations. For spherical crystallites, the magic numbers represent the number of atoms necessary to fill in symmetric geometric forms, leading to isocohedral clusters, which then start to adapt the final crystal structure.

When the concentration is closer to the supersaturation limit, heterogeneousnucleation occurs most often. The nucleus develops onto the substrate, with which it makes a contact angle α. Solution of the equations for the nucleus size and activation energy imply that the critical radius is the same as for homonuclear nucleation. On the other hand, the activation energy for heteronuclear nucleation is the product of the activation energy for homogeneous nucleation and a term depending on the value of α, since

$$\Delta G_{het}^{*} = \Delta G^{*}\left(\frac{1}{2} - \frac{3}{4}\cos\alpha + \frac{1}{4}\cos^{3}\alpha\right) \tag{4.14}$$

The influence of the contact angle, α, is worth demonstrating. For example, if α is 180°, the term in parentheses is 1 and $\Delta G_{het} = \Delta G^{*}$. If α is 90°, $\Delta G_{het} = \Delta G^{*}/2$, and so on. The smaller the angle, the greater the contact the substrate has with the supersaturated solution, and the lower the energy required for nucleation.

While thermodynamics dictates whether or not nucleus formation occurs, thenucleationrate is often the more important factor. The nucleation rate, J, can be defined as the product of a kinetic factor and the activation energy:

$$J = K_0 \exp\left(\frac{-\Delta G^{*}}{k_B T}\right) \tag{4.15}$$

In this case, the kinetic factor is dependent on the solubility of the materials and the frequency at which critical nuclei become supersaturated and transform into crystals. Hence,

$$K_0 = N_0 v_0 \tag{4.16}$$

From these last two equations, it becomes clear that all other factors being equal, homogeneous nucleation is more rapid in solutions where the solubility is high. If the solubility is high, the interfacial energy (the cost associated with nucleus formation) between the crystal and the solution is low. In the opposite situation, when the material is sparingly soluble, nucleation is more difficult and occurs only with a high supersaturation concentration. In this situation, nucleation most often is a catastrophic situation, leading to smaller, poor-quality crystallites. Often in small systems it is very important to tailor the concentration carefully around the supersaturation limit. Although the nucleation rate is strongly dependent on the concentration, it also depends on the way in which the high concentration was achieved and the rate at which it was approached.

Eventually, upon continued growth, stable clusters begin to adapt the final crystalform. Gibbs demonstrated that in the case of growing crystals there is a condition

$$\sum \sigma_i F_i \tag{4.17}$$

which is a minimum. In this condition, the subscript i represents the ith face of acrystal and F is the surface area of the crystal, while σ is the specific free energy. Von Laue's work allows us to rephrase this condition and say that.

In general, as surface energy increases, the area of the crystal face decreases. This condition has undergone several adaptations over the years, first by Bornin his lattice theory. In modern crystal theory, what is more important is not the direct relationship between the surface free energy, but instead, the energy released when a building block is attached to the surface of a growing crystal. The principal consequence of this adaptation is that, as attachment energy increases, the energy required for a bond to form decreases. As a result, the displacement velocity, or the rate of filling of the face, also increases. To understand this further, let's examine a 2D crystal. In our first crystal (Fig. 4. 12(a)), there are two different types of bonds: a in the [10] direction and b in the [01]. Now assume that the attachment energy resulting from formation of a bonds is greater than that of b. This would result in the [10] face having a greater displacement velocity (i. e., growing faster than the [01] face). In our second example (Fig. 4. 12(b)), there are two types of bonds, a strong A—B bond, and a weaker A—B′ bond along the [10] face, and then the a bonds on the [01] face are equal to the weaker A—B bonds.

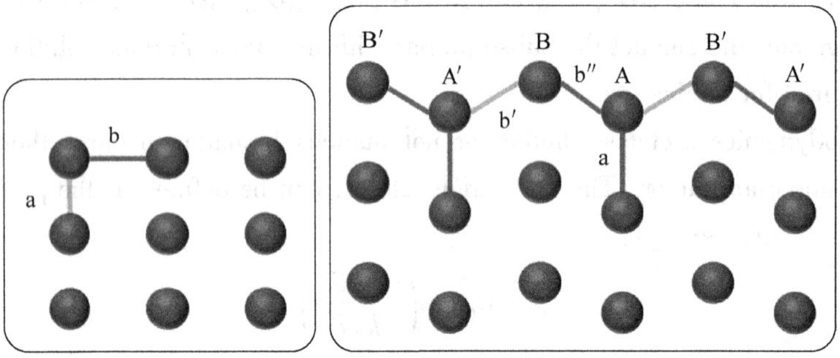

Fig. 4. 12 (a) Two-dimensional crystal representation in which bond a is stronger than bond b; (b) Two-dimensional crystal representation in which bonds a and b are the same strength while b is much weaker

Now the growth is governed by the formation of two bonds in the [01] direction, one stronger and one weaker, compared to two weaker bonds in the [10] direction. The displacement velocity in the [01] direction is, however, governed by the weaker bond length. Consequently, the crystal grows nearly equally in both the [01] and [10] directions, which leads to two observations. First, for a crystal to grow in the direction of a strong bond, these bonds must be uniform and form an uninterrupted chain. If the bond chain contains bonds of different types, the influence on the shape is determined by the weakest bond present in the chain.

These observations were made and explained by P. Hartman and W. G. Perdokin their periodic bond chain vector (PBC) theory. This theory takes into account the fact that 3D crystals are far more complicated structures, which are full of partial bonds and preferred directions. When the PBC vector is parallel with a crystal face, there is maximum growth along that crystal face. Using PBC theory, the faces of a growing crystal can be divided into three categories: flat faces (class F), stepped faces (class S), and kinked faces (class K). In Fig. 4. 13, there are three PBC vectors, which define six class F faces. Any face that is parallel to one of the A, B, C vectors is a class S face. Faces that are not parallel to any of the vectors become class K faces. As a crystal grows along the F face, a relatively small amount of attachment energy is released as the building block is attached to the crystal face. For class S, two stronger bonds are formed as the building block is attached, whereas in class F, three bonds are formed. This results in, acceleration during the growth; the displacement velocity of class K is faster than that of class S, which is faster than that of class F. As the class K faces get filled in more quickly, they become class S faces, which, in turn, become F faces in the final crystal. Class K faces are rarely seen in the final crystal structure. It is important to note here that we are talking about the final crystallization thermodynamics. In the initial stages, the growth is governed nucleation. Specifically, the crystallite germ grows initially in all directions, resulting in an isometric crystallite. As the crystallite grows into a bigger crystal, the faces start to emerge and the final product is governed by the attachment energy.

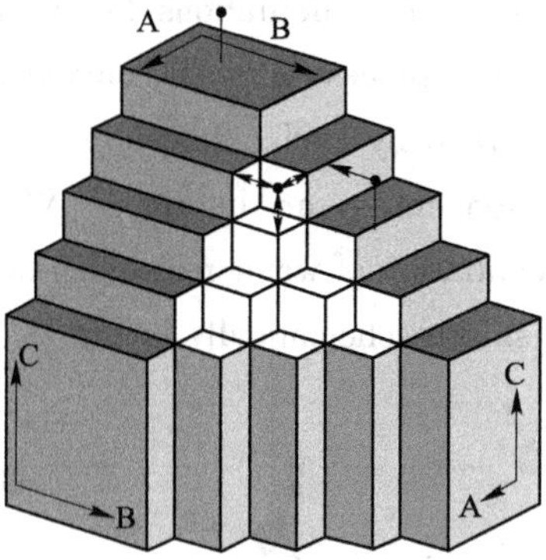

Fig. 4. 13 Three-dimensional crystal with three periodic bond chain vectors
(The figure shows F-faces (100), (010), and (001); S-faces (110), (101), and (011); and the K-face (111))

In the case of a perfect crystal (i. e., a crystal where there are no defects in a crystal face), the growth takes place by two-dimensional nucleation. To create the 2D nucleus, building blocks just adsorb on the faces and then diffuse together and cluster. Once the nucleus has reached critical size, it becomes stable and exhibits some class K faces, where growth units can be incorporated that convert the class K face to class S. The process continues until all the K faces are filled; the S sites then begin to fill, resulting in a perfect crystal that has all F faces. The kinetics of the growth is typically diffusion-controlled at the rate R:

$$R = J_2 dS \tag{4.18}$$

It is equal to the product of the 2D nucleation rate (J), the height of the layer (d) and area(S) of the face.

4.4 Gas State Reaction

In this chapter we examine the interactions that occur between a solid and a vapor. These interactions can lead to intercalation, physical vapor deposition, or chemical vapor deposition. The latter two interactions, along with molecular beam epitaxy, which will be discussed briefly, are used to produce thin films on substrates. The production of thin films has wide applications encompassing integrated-circuit manufacturing, optical devices, chemical sensors, and ionic conductors to name a few.

We first review the properties of gases with respect to the important relationships needed when considering film growth. Next, we review the isotherms typically used when considering the adsorption of gases on solid surfaces. After reviewing gases and adsorption, we examine intercalation, physical vapor deposition, chemical vapor deposition, and molecular beam epitaxy. In this chapter we refer to the vapor-phase species as molecules even if the actual species consists of monatomic atoms such as $O_2(g)$ or $Ar(g)$.

4.4.1 Vapor Absorption and Adsorption

We begin by defining two important terms: absorption and adsorption. The International Union of Pure and Applied Chemistry offers precise definitions for absorption and adsorption (IUPAC, 1972). *Absorption* is used to describe aprocess where a component is transferred from one phase to another. Hydrogen gas can be absorbed by $LaNi_5$. *Adsorption* is used to describe the increased or decreased concentration of a component at an interface. Water molecules will adsorb to an aluminum oxide surface. The concentration of water molecules on the aluminum oxide surface will be greater than the vapor phase, shown schematically in Fig. 4. 14.

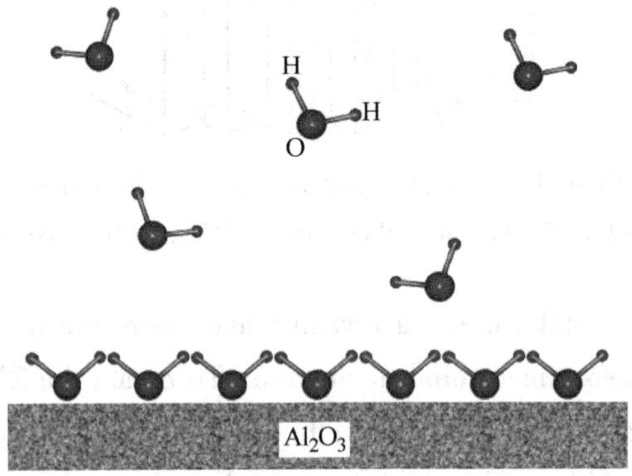

Fig. 4. 14 Water molecules adsorbing on a surface of Al_2O_3 where the surface concentration of water is higher than the bulk vapor concentration

Absorption, in the context of a vapor-solid interface, means that vapor molecules penetrate the solid to occupy vacancies or interstices within the interior of the solid. Adsorption, in the same context, means that vapor molecules stick to the surface of a solid but do not migrate to the interior of the solid. This distinction is easier to apply to a single-crystal surface where there is a clear delineation of surface and interior atoms. In a high-surface-area solid such as an aerogel, which can have a surface area as high as $3000m^2/g$, such distinctions become problematic. Both adsorption and absorption begin with adsorption, although in some cases an adsorbed molecule may be in such a state for only a brief time before moving to the interior.

Because any surface deposition process begins with adsorption, we begin our discussion there. Molecules adsorb on the surface of a solid for a variety of reasons, and the extent of adsorption depends on several factors, some of which can be controlled by the synthetic chemist. The primary reason that a molecule adsorbs to the surface of a solid is the exothermic nature of chemical bond formation between the substrate atoms and the vapor-phase molecules. Adsorbed molecules can bond to a surface atom via a covalent bond or London (Vander Waals) forces. Those that adsorb due to covalent bonding with the surface are referred to as *chemisorbed*, and those that adsorb due to London forces are referred to *physisorbed*. The strength of the bonding interaction will, inpart, determines the mobility of the adsorbed atom. This mobility will affect the subsequent fate of the adsorbed atom with respect to intercalation, chemical vapordeposition, or physical vapor deposition.

The nature of surface adsorption has traditionally be characterized with the construction of a variety of isotherms, the two most common being the Langmuir isotherm and the Brunauer-Emmett-Teller (BET) isotherm. Most physical chemistry undergraduate texts present a detailed explanation of at least these two isotherms. An *isotherm* is a mathematical relationship relating how much of a surface is covered with adsorbed molecules as a function of the pressure of the gas consisting of those molecules. As the name implies the relationship is for a given temperature. Intuitively, one would expect the surface coverage to increase with increasing pressure, but the microscopic details of how the rates of adsorption and desorption will affect the surface coverage must also be taken into account.

The *Langmuir isotherm*, regarded as the starting point for any discussion of isotherms because of its simplicity and applicability, was derived by Irving Langmuir in 1918. Langmuir made four assumptions: (1) The adsorbed molecules do not interact with each other. (2) The energy released when a molecule binds to the surface is independent of how much of the surface is covered with adsorbed molecules. (3) A fixed number of sites are available for molecular adsorption. (4) Only one layer of adsorbed molecules will form.

A gas, X(g), at a given pressure is assumed to be in equilibrium with adsorbed molecule, X(ads):

$$X(g) + \text{surface} \xrightarrow{k_1} X(\text{ads}) : \text{surface}$$

$$X(\text{ads}) : \text{surface} \xrightarrow{k_{-1}} X(g) + \text{surface}$$

where k_1 is the rate constant for adsorption and k_{-1} is the rate constant for desorption. The fixed number of sites available for molecular adsorption is N_0. The fraction of these sites occupied at equilibrium is traditionally labeled θ. The rate of desorption will be equal to

$$\text{rate} = k_{-1}\theta N_0 \tag{4.19}$$

and the rate of adsorption will be

$$\text{rate} = k_1 P(1 - \theta) N_0 \tag{4.20}$$

where P is the pressure of the gas. The adsorption rate is linear in P because it will depend on the collision frequency of the gas with the surface, and this collision frequency is linear in pressure. Equating these rates at equilibrium and solving for θ results in

$$\theta = \frac{aP}{1 + aP} \tag{4.21}$$

where $a = k_1/k_{-1}$. This is the desired result, a functional form relating surface coverage, θ, to equilibrium pressure, P. The temperature dependence of θ is contained in the temperature dependence of k_1 and k_{-1}. The Langmuir isotherm provides important physical insights. First, at low pressure, the surface coverage will be linear in pressure. Second, at high pressure, all the adsorption sites will be filled and no additional molecules can be adsorbed. Experimentally, the Langmuir isotherm is investigated by a series of well-controlled steps. First, the surface to be investigated is cleaned of adsorbed species, typically by heating it. Second, a known amount of gas in introduced into the sample chamber holding the sample (Fig. 4. 15 (a)). Because gas pressures are low, the ideal gas law can be used to calculate the number of moles of gas introduced given P_0, V, and T. The equilibrium P then allows one to calculate the amount of gas adsorbed (Fig. 4. 15(b)). A series of gas injections of increasing amount of gas allows one to produce a plot of adsorbed molecules versus equilibrium pressure. This is an isotherm. Using the ideal gas law, these adsorbed molecules, even though they are not in the gas phase, can be reported as the equivalent volume, v, of gas at 0℃ and 101. 325kPa, standard temperature and standard pressure (STP) in the gas world. Often, then, it is not the number of moles of gas adsorbed that is plotted in an isotherm, but rather, this equivalent volume.

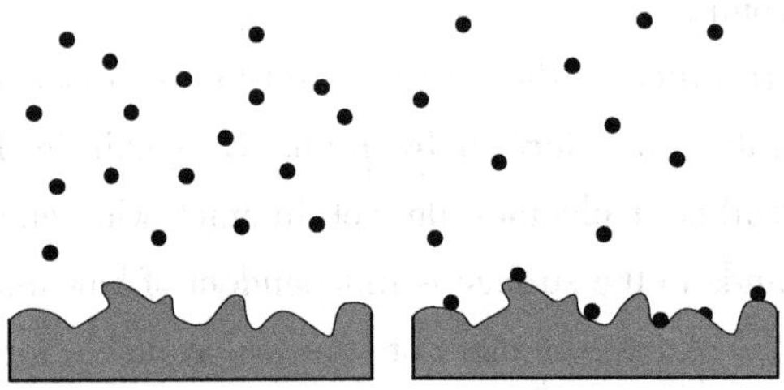

Fig. 4. 15 (a) Clean surface and gas prior to adsorption; (b) equilibrium between adsorbed gas and vapor-phase gas

Most of the assumptions in deriving the Langmuir isotherm are not accurate. It works best for cases of chemisorption where only one layer is formed. A straight forward check on the validity of the Langmuir isotherm is to plot the inverse of the surface coverage, θ, versus the inverse of the equilibrium pressure:

$$\frac{1}{\theta} = \frac{1}{aP} + 1 \tag{4.22}$$

An alternative and equivalent verification can be sought by using the equivalent volume, v, mentioned above instead of the surface coverage. The surface coverage equals the ratio of the equivalent volume to the equivalent volume of a complete monolayer,

$$\theta = \frac{V(P)}{V(P \to \infty)} \tag{4.23}$$

This results in a verification equation of

$$\frac{1}{V(P)} = \frac{1}{aPV(P \to \infty)} + \frac{1}{V(P \to \infty)} \tag{4.24}$$

Isotherms that follow the Langmuir isotherm (i. e, form only a monolayer) are termed type Ⅰ isotherms. The BET isotherm addresses some of the shortcomings of the Langmuir isotherm: most important, the restriction of one adsorbed layer. The resulting isotherm is

$$\frac{P}{(P^{*}-P)V}=\frac{1}{V_{mon}b}+\frac{b-1}{V_{mon}b}\frac{P}{P^{*}} \tag{4.25}$$

where b is a constant at a given temperature; V_{mon} is the equivalent volume of a monolayer of adsorbed gas; P^{*} is the equilibrium vapor pressure of the liquid phase of the adsorbed gas at the experimental temperature. A plot of $P/(P^{*}-P)V$ versus P/P^{*} is constructed with the linear portion giving b and, more important, V_{mon}. When multiple adsorbed layers are formed, the isotherm is called a *type Ⅱ isotherm*. The BET isotherm is an example of a type Ⅱ isotherm. Type Ⅰ and type Ⅱ isotherms have characteristic shapes as shown in Fig. 4.16, where the equivalent volume of adsorbed gas is plotted versus the equilibrium vapor pressure. Type Ⅰ (Fig. 4.16(a)) reaches a plateau when the first monolayer is completed and no subsequent monolayers form. Type Ⅱ (Fig. 4.16(b)), where the BET isotherm is typically applied; shows an infection point where the first monolayer is completed and subsequent layers are beginning to form. In both of these isotherms, the equilibrium pressure, P, is typically much less than the equilibrium vapor pressure of the liquefied gas, P^{*}.

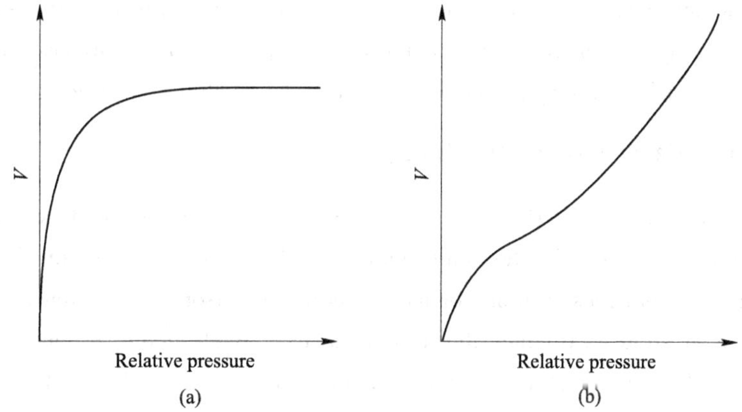

Fig. 4.16 Isotherms
(a) Type Ⅰ; (b) Type Ⅱ

Both isotherms contain a term representing the volume of a monolayer of adsorbed gas that, again using the ideal gas law, can be converted to the number of molecules needed to complete a monolayer. This has several practical consequences. First, if the size of the molecule is known, the surface area of the solid can be determined. Several gases are used routinely for this purpose: nitrogen, argon, krypton, and oxygen. The accepted surface areas of these molecules are: nitrogen ($0.162nm^2$), krypton ($0.195nm^2$), argon ($0.138nm^2$), and oxygen ($0.141nm^2$).

Alternatively, if the surface area is known, an estimate of molecule size is possible. In our discussion of physical and chemical vapor deposition we make use of the Langmuir isotherm because of the low pressures involved. As already stated, the strength of the bonding between the substrate and the adsorbed species distinguishes physical adsorption from chemical adsorption. The strength of this bonding, termed the *enthalpy of adsorption*, can be determined using

$$\left(\frac{\partial \ln P}{\partial T}\right)_{\theta}=-\frac{\Delta H_a}{RT^2} \tag{4.26}$$

where ΔH_a is the differential molar enthalpy of adsorption; P is the pressure of the gas in equilibrium with the solid surface; T is the temperature; R is the gas constant. The enthalpy of adsorption for a physisorbed species will typically be less than 20kJ/mol (exothermic), whereas that of a chemisorbed species will typically be greater than 200kJ/mol (exothermic). To extract ΔH_a from Eq. (4.25), a series of isotherms at different temperatures is collected. The coldest corresponds to the upper trace in Fig. 4.17, and the warmest to the lower trace. The P and T data points at constant θ are indicated with the tie line. It is interesting to consider that adsorption is almost always an exothermic process because of the formation of a chemical bond, regardless of whether covalent or van der Waals in character, yet the processes we discuss presently, particularly chemical vapor deposition, typically use surfaces with very high temperatures where surface adsorption is not favored by Le Chatelier's principle.

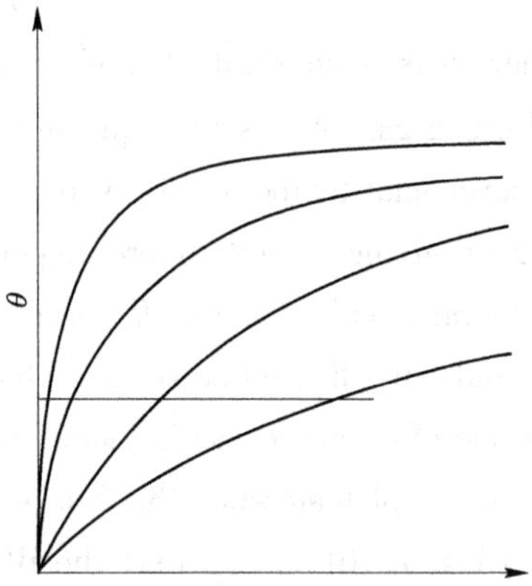

Fig. 4.17 Series of isotherms needed to determine ΔH_a

4.4.2 Chemical Vapor Deposition (CVD)

Chemical vapor deposition (CVD) is a process commonly used to produce high-purity solid materials. The process is used in the semiconductor industry to produce thin films. In a CVD process, the substrate is exposed to one or more volatile precursors, which react or decompose on the surface of the substrate to produce the desired deposit. Such atomistic deposition can provide highly pure materials with structural control at atomic or nanometric level. It can also produce single-layer and multilayer materials, composites, nanostructured materials etc. with controlled dimensions and unique structures at fairly low temperatures. The versatility of CVD has made it one of the main methods for the deposition of thin films and coatings of semiconductors (e. g. Si, Ge, $Si_{1-x}Ge_x$, Ⅲ-Ⅴ, Ⅱ-Ⅵ), dielectrics (e. g. SiO_2, AlN, Si_3N_4), ceramic materials (e. g. SiC, TiN, TiB_2, Al_2O_3, BN, $MoSi_2$, ZrO_2), oxidation or diffusion barriers, metallic films (e. g. W, Mo, Al, Au, Cu, Pt), fibers (e. g. B and SiC monofilament fibers) and fiber coatings. CVD is practised in a variety of formats. CVD processes differ in the methods by which the chemical reactions are initiated. For example, they can be thermally activated, plasma-enhanced, photo-assisted, or metal-organic-assisted. In this chapter, we will give examples of the synthesis of some of the important inorganic materials by CVD.

Polycrystalline Si (poly-Si) thin films are used in integrated circuits as gate electrodes, emitters in bipolar transistors, load resistors and interconnection connectors. SiH_4 is generally used as the precursor, where it undergoes pyrolysis at reduced pressure (typically ~133Pa) in the presence of H_2, He or N_2 to deposit poly-Si at temperatures between 610℃ and 630℃. Epitaxial Si films are deposited using SiH_4, $SiCl_4$ and $SiCl_2H_2$, which undergo decomposition (i.e. pyrolysis), reduction or disproportionation. reduction of $SiCl_4$ in hydrogen has been widely used for homoepitaxial growth of Si films.

The common gases used for CVD synthesis of SiO_2 are silane and oxygen, dichlorosilane ($SiCl_2H_2$) and nitrous oxide (N_2O), or tetraethylorthosilicate ($Si(OC_2H_5)_4$). Si_3N_4 has high electrical resistance and dielectric strength and is suitable as a passivating layer and storage capacitor in dynamic random access memory. The Si source for Si_3N_4 can be $SiCl_4$, or $SiCl_2H_2$, the nitride source being NH_3 or N_2+H_2. The common precursor for the CVD of Si_3N_4 in the semiconductor industry is $SiCl_2H_2+NH_3$ and the deposition is operated at a temperature between 750℃ and 900°C and a pressure of 25~115Pa.

Ⅲ-Ⅴ nitrides (InN, GaN and AlN) are made by CVD using $(CH_3)_3M$ (M = In, Ga and Al) and NH_3 as precursors. HgCdTe, which is used in infrared (Ir) detectors, is made by the reaction of the dimethyl derivatives of the metals. PbSnTe is also synthesized by CVD. High-quality thin films of superconducting $yBa_2Cu_3O_{7-x}$ have been prepared by metal organic chemical vapour deposition (MOCVD) techniques using β-diketonate derivatives of y, Ba and Cu as precursors.

CVD has been recently used to produce bulk quantities of graphene and N-doped graphene. Solutions containing ferrocene and thiophene in ethanol have been used for the synthesis of graphene. An aerosol was generated ultrasonically and then carried by an argon flow into a quartz tube located inside a two-furnace system heated to 1223K. After completion of the reaction, the quartz tube was removed and black graphene powder was found on the walls of the tube located in the first furnace area. Graphene is deposited on Ni, Co or Cu foils or films of CVD based on CH_4 or C_2H_4. Fullerene-like MoS_2 and atomic layer-thick MoS_2 have also been synthesized by CVD.

4.4.3 Atomic Layer Deposition (ALD)

Atomic layer deposition (ALD) is a self-limiting (the amount of film material deposited in each reaction cycle is constant), sequential surface chemistry technique that deposits thin films of materials on solid substrates. Fig. 4.8 shows a schematic diagram of the sequential, self-limiting surface reactions during ALD. Due to the nature of these reactions, ALD film growth makes it possible to have atomic scale deposition control. The chemistry of ALD is similar to that of CVD, except that the reaction in ALD breaks the CVD reaction into two half-reactions, keeping the precursor materials separate. By keeping the precursors separate throughout the deposition process, atomic-layer control of film growth can be obtained as fine as ~0.01nm (10pm) per cycle. Separation of the precursors is accomplished by pulsing a purge gas (typically nitrogen or argon) after each precursor pulse to remove the excess precursor from the reaction chamber. This avoids parasitic CVD deposition on the substrate. We provide a few examples of important ALD reactions.

ALD of Al_2O_3 is a model system. ALD of Al_2O_3 is usually performed using trimethylaluminum (TMA) and H_2O. The surface chemistry can be described as follows:

$$AlOH^* + Al(CH_3)_3 \longrightarrow AlOAl(CH_3)_2^* + CH_4$$
$$AlCH_3^* + H_2O \longrightarrow AlOH^* + CH_4$$

Here, the asterisks denote surface species. Al_2O_3 ALD growth occurs during alternating exposures to TMA and H_2O. Al_2O_3 ALD is a model system because the surface reactions are efficient and self-limiting. The main driver of the efficient reactions is the formation of strong Al—O bonds. The overall reaction for Al_2O_3 ALD is given by

$$2Al(CH_3)_3 + 3H_2O \longrightarrow Al_2O_3 + 3CH_4 \quad \Delta H = -376kJ/mol$$

This reaction has an extremely high reaction enthalpy.

Questions

1. Select one or more of the following modern items or devices and conduct an Internet search in order to determine what specific material (s) is (are) used and what specific properties this (these) material(s) possess(es) in order to the device/item to function properly. Finally, write a short essay in which you report your findings. For examples: Cell phone/digital camera batteries; Cell phone displays; Solar cells; Wind turbine blades; Fuel cells; Automobile engine blocks.
2. Briefly cite the main differences between metals, ceramics, and polymeric materials.
3. What type (s) of bonding would be expected for each of the following materials: brass (a copper-zinc alloy), rubber, barium sulfide (BaS), solid xenon, bronze, nylon, and aluminum phosphide (AlP).
4. What is the definition of solid state reaction? Please illustrate typical features of solid state reaction.
5. Please make classic examples to describe the evolution of solid-solid interfaces in the solid-solid reaction.
6. Explain the detailed procedure of Ostwald Ripening and make several examples.
7. What is the difference between chemical vapor deposition (CVD) and atomic layer deposition (ALD)?
8. What is the importance of vapor absorption and adsorption in the gas state reaction?

References

[1] Ashby M F, Jones D. Engineering materials 1: an introduction to properties, applications and design [J]. Third Edition. Fuel and energy abstracts, 2011,36 (4):284.

[2] Ashby M F, Jones D. Engineering materials 2: An introduction to microstructures, processing, and design [M]. Cham:Spring, 2013.

[3] Ashby M, Sherdiff H, Cebon D. Materials: Engineering, science, processing and design[J]. Materials Today,2007,10 (5):59.

[4] Brady, James E. Chemistry : matter and its changes[M]. Hoboken:Wiley, 2009.

[5] Schaffer, James P. The science and design of engineering materials[M]. 2nd edition. New York:McGraw-Hill College, 2003.

[6] Shackelford J F. Introduction to materials science for engineers[J]. Pearson Schweiz Ag,2000,22 (1):

51~93.

[7] Cave G W V, Raston C L, Scott J L. Recent advances in solventless organic reactions: towards benign synthesis with remarkable versatility[J]. Chemical Communications,2001(21):2159~2169.

[8] Rothenberg G, Downie A P, Raston C L, et al. Understanding solid/solid organic reactions[J]. Journal of the American Chemical Society, 2001,123 (36):8701~8708.

[9] Braga D, Grepioni F. Reactions between or within molecular crystals[J]. Angewandte Chemie International Edition,2004,43 (31):4002~4011.

[10] Ohhara T, Harada J, Ohashi Y, et al. Direct observation of deuterium migration in crystalline-state reaction by single-crystal neutron diffraction. Ⅱ. 3-1 Photoisomerization of a cobaloxime complex [J]. Acta Crystallographica Section B: Structural Science, 2000,56 (2):245~253.

[11] Luty T, Eckhardt C J. General theoretical concepts for solid state reactions: quantitative rormulation of the reaction cavity, steric compression, and reaction-induced stress using an elastic multipole representation of chemical pressure[J]. Journal of the American Chemical Society,1995,117 (9):2441~2452.

[12] Meth-Cohn O, Wang M X. Regioselective biotransformations of dinitriles using rhodococcus sp. AJ270 1 [J]. Journal of the Chemical Society, Perkin Transactions 1, 1997(21):3197~3204.

[13] Cohen M D. The photochemistry of organic solids[J]. Angewandte Chemie International Edition in English, 1975,14 (6):386~393.

[14] Ohashi Y. Dynamical structure analysis of crystalline-state racemization [J]. Accounts of Chemical Research, 1988,21 (7):268~274.

[15] Wilkinson G, Gillard R D, McCleverty J A. Comprehensive coordination chemistry: the synthesis, reactions, properties and applications of coordination compounds[M]. Oxford: Pergamon Press,1987.

[16] Evans S V, Garcia-Garibay M, Omkaram N, et al. Use of chiral single crystals to convert achiral reactants to chiral products in high optical yield: application to the di-. pi. -methane and Norrish type Ⅱ photorearrangements[J]. Journal of the American Chemical Society, 1986,108 (18): 5648~5650.

[17] Kumar P J, Nishimura K, Senna M, et al. Novel low-temperature solid-state route for nanostructured cubic garnet LiLaZrO and its application to Li-ion battery[J]. RSC Advances, 2016, 6: 62656.

[18] Gerd Kaupp . Solid-state molecular syntheses: complete reactions without auxiliaries based on the new solid-state mechanism[J]. Cryst. Eng. Comm. ,2003,5 (23): 117~133.

[19] Dittmer D C. 'No-solvent' organic synthesis[M]. Hoboken:Wiley, 1997.

[20] Cave G, Raston C L, Scott J L. Recent advances in solventless organic reactions: towards benign synthesis with remarkable versatility[J]. Chemical Communications, 2001,21 (21):2159~2169.

[21] Toda F. Solid state organic chemistry: efficient reactions, remarkable yields, and stereoselectivity [J]. Cheminform,1996,27(16):480~486.

[22] Ramamurthy V,Venkatesan K. Photochemical reactions of organic crystals[J]. Chemical Reviews, 1987, 87 (2):433~481.

[23] Garcia-Garibay M A. Chemical reactivity in organized media: statistical entropy and information in crystals and enzymes[J]. Current Opinion in Solid State and Materials Science, 1998,3 (4):399~406.

[24] Ohashi Y. Crystalline-state photoreactions: direct observation of reaction processes and metastable intermediates[M]. Springer Japan,2014.

[25] Braga D, Desiraju G R, Miller J S,et al. Innovation in crystal engineering[J]. Crystengcomm, 2002, 4 (83):500~509.

[26] Maginn S J. Crystal engineering : the design of organic solids by G. R. Desiraju[J]. Journal of Applied Crystallography, 1991,24 (3): 265.

[27] Fisler D K, Cygan R T. Diffusion of Ca and Mg in calcite[J]. American Mineralogist, 1999, 84: 1392~1399.

[28] John N. Lalena, David A. Cleary, Everett Carpenter, et al. In inorganic materials synthesis and fabrication [M]. Hoboken: Wiley-Interscience, 2008.

[29] Bernal J D. The Bakerian Lecture, 1962. The structure of liquids[J]. Proceedings of the Royal Society of London. Series A. Mathematical and Physical Sciences, 1964,280 (1382):299~322.

[30] Billinge S J L. How do your crystals grow? [J]. Nature Physics, 2009,5 (1):13~14.

[31] Madras G, McCoy B J. Ostwald ripening with size-dependent rates: similarity and power-law solutions[J]. The Journal of Chemical Physics, 2002,117 (17):8042~8049.

[32] Stoica C, Tinnemans P, Meekes H, et al. Epitaxial 2D nucleation of metastable polymorphs: a 2D version of Ostwald's rule of stages[J]. Crystal Growth & Design, 2005,5 (3):975~981.

[33] Nývlt J. The Ostwald rule of stages[J]. Crystal Research and Technology, 1995,30 (4): 443~449.

[34] Antonietti M, Kuang D, Smarsly B, et al. Ionic liquids for the convenient synthesis of functional nanoparticles and other inorganic nanostructures[J]. Angewandte Chemie International Edition, 2004, 43 (38): 4988~4992.

[35] Baldan A. Review progress in Ostwald ripening theories and their applications to nickel-base superalloys part Ⅰ: Ostwald ripening theories[J]. Journal of Materials Science, 2002,37 (11):2171~2202.

[36] Klausner J F, Li Y, Darwish M, et al. Innovative diffusion driven desalination process[J]. Journal of Energy Resources Technology, 2004,126 (3): 219~225.

[37] Boistelle R, Astier J P. Crystallization mechanisms in solution[J]. Journal of Crystal Growth, 1988,90: 14~30.

[38] Echt O, Sattler K, Recknagel E. Magic numbers for sphere packings: experimental verification in free xenon clusters[J]. Physical Review Letters, 1981,47 (16):1121~1124.

[39] Wills J M. Lattice packing of spheres and the Wulff-shape[J]. Mathematika, 1996,43 (2):229~236.

[40] Hartman P, Chan H K. Application of the periodic bond chain (PBC) theory and attachment energy consideration to derive the crystal morphology of hexamethylmelamine[J]. Pharmaceutical Research, 1993, 10 (7):1052~1058.

[41] Van Mal H H, Buschow K H J, Miedema A R. Hydrogen absorption in $LaNi_5$ and related compounds: experimental observations and their explanation[J]. Journal of the Less Common Metals, 1974,35 (1): 65~76.

[42] Al-Abadleh H A, Grassian V. FT-IR study of water adsorption on aluminum oxide surfaces[J]. Langmuir, 2003,19:341~347.

[43] Azizian S, Eris S, Wilson L D. Re-evaluation of the century-old langmuir isotherm for modeling adsorption phenomena in solution[J]. Chemical Physics, 2018,513:99~104.

[44] Sing K S W. Reporting physisorption data for gas/solid systems with pecial reference to the determination of surface area and porosity (Recommendations 1984) [J]. Pure and Applied Chemistry, 1985,57 (4): 603~619.

[45] Bromberg, Philip J. Physical chemistry[M]. 2nd edition. Boston: Allyn & Bacon, 1984.

[46] Choy K L. Chemical vapour deposition of coatings[J]. Progress in Materials Science, 2003,48 (2): 57~170.

[47] Van Den Brekel C H J, Bollen L J M. Low pressure deposition of polycrystalline silicon from silane[J]. Journal of Crystal Growth, 1981,54 (2):310~322.

[48] Hitchman M L, Kane J, Widmer A E. Polysilicon growth kinetics in a low pressure chemical vapour

deposition reactor[J]. Thin Solid Films,1979,59 (2): 231~247.

[49] Liaw H M, Rose J W. 1 - Silicon vapor-phase epitaxy[M]. In Epitaxial Silicon Technology, Baliga, B. J. , Ed. Academic Press: 1986.

[50] Schuegraf K. Handbook of thin-film deposition processes and techniques: principles, methods, equipment, and appl[M]. Park Ridge:Noyes Publication,1988.

[51] Egawa T, Ishikawa H, Jimbo T, et al. Heteroepitaxial growth of Ⅲ-Ⅴ compound semiconductors for optoelectronic devices[J]. Bulletin of Materials Science, 1999,22 (3): 363~367.

[52] Anderson P L, Erbil A, Nelson C R, et al. A high-speed, rotating-disc metalorganic chemical vapor deposition system for the growth of (Hg, Cd)Te and related alloys[J]. Journal of Crystal Growth, 1994, 135 (3): 383~400.

[53] Manasevit H M, Ruth R P, Simpson W I. Properties of PbTe and $Pb_{1-x}Sn_xTe$ films prepared by metalorganic chemical vapor deposition[J]. Journal of Crystal Growth, 1986,77 (1):468~474.

[54] Berry A D, Gaskill D K, Holm R T, et al. Formation of high Tc superconducting films by organometallic chemical vapor deposition[J]. Applied Physics Letters, 1988,52 (20):1743~1745.

[55] Campos-Delgado J, Romo-Herrera J M, Jia X, et al. Production of a new form of sp2 carbon: crystalline graphene nanoribbons[J]. Nano Letters, 2008,8 (9): 2773~2778.

[56] Wei D, Liu Y, Wang Y, et al. Synthesis of N-doped graphene by chemical vapor deposition and its electrical properties[J]. Nano Letters,2009,9 (5):1752~1758.

[57] Etzkorn J, Therese H A, Rocker F, et al. Metal-organic chemical vapor depostion synthesis of hollow inorganic-fullerene-type MoS_2 and $MoSe_2$ nanoparticles [J]. Advanced Materials, 2005, 17 (19): 2372~2375.

[58] Lee Y H, Zhang X Q, Zhang W, et al. Synthesis of large-area MoS_2 atomic layers with chemical vapor deposition[J]. Advanced Materials, 2012,24 (17):2320~2325.

[59] George S M. Atomic layer deposition: an overview[J]. Chemical Reviews, 2010,110 (1):111~131.

[60] George S M, Ott A W, Klaus J W. Surface chemistry for atomic layer growth[J]. The Journal of Physical Chemistry, 1996,100 (31):13121~13131.

Chapter 5 Preparation Methods of Metal

Powder metallurgy is a process for preparing metal powders and using metal (or metal and non-metal mixture) powder as raw materials to form parts and products by molding and sintering. As the main raw material of the industry, metal powder is widely used in the fields of machinery, metallurgy, chemical industry, and aerospace materials. Metal powder is the basic raw material of the powder metallurgy industry, and its output and quality determine the development of the powder metallurgy industry. In this article, we will take a close look at the preparation methods of metal powder. The metal powder is usually an aggregate of metal particles of less than 1mm, and there is no uniform regulation of the division of the particle size interval. Each of the powder particles may be a crystal or a plurality of crystals depending on the particle size and preparation method.

In recent years, the increasing number of published papers shows great interest of engineers and scientists for the importance of metal nanoparticles (NPs) such as silver (Ag), gold (Au) and copper (Cu) etc., because of their physicochemical properties, nano meter (nm) size and surface plasmon behavior. These metal NPs are used in catalysis, electronics for circuit development and as self-assembled new nanostructure materials, but only show different sets of applications with light interaction, called as plasmonics. The self-assembly of metal nanoparticles into small clusters, provides suitability to control and tune their optical properties by coupling surface plasmon of nanoparticles and enhance the electric field in between the particles gaps. These gaps in self-assembled nanostructures with high intensity field are known as hot spots. The enhancement of electric field is very important in terms of optical and chemical properties of nanoparticles as plasmonic properties are strongly influenced by electric field. Various methods are adopted for the preparation of metal nanostructures that should be ideally controllable regarding shape and size, cheap, environment friendly, and high product yield with less waste products.

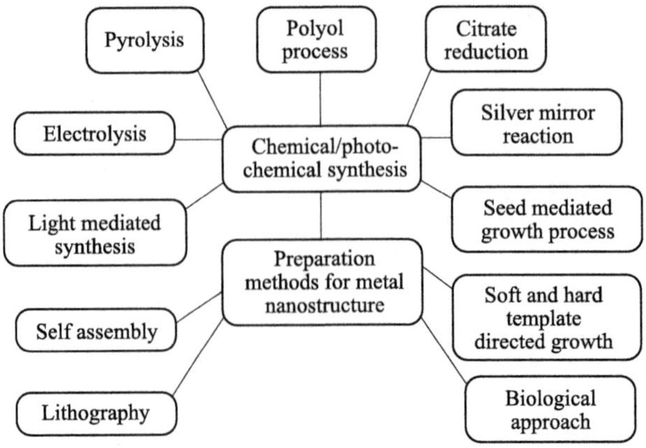

Fig. 5. 1 Various methods for synthesis of metal nanostructures

Fig. 5.1 summarizes various preparation methods of metal nanostructures. Here, we will discuss all synthetic processes in reference mostly of Ag, as an example.

5.1 Chemical and Photochemical Synthesis

The optical properties of metal nanostructures are highly dependent on their size and shape. The controlled morphology of nanostructures is an important factor which has strong impacts on their unique plasmonic properties. Until now, various methods are adopted for the preparation of controlled size and shape of nanostructures. In general, metal/Ag^+ salt precursor is mixed with reducing agent in the presence of stabilizing agentand control the size and shape of metal nanostructures. Past research publications show that silver nitrate, as a precursor, has been abundantly used due to its cheap price and easy availability. Various reducing agents such as sodium citrate, sodium borohydride and alcohols are widely used to reduce metal/Ag^+ ions present in solution into metal/Ag atoms which combine to make aggregates and finally become nanostructures. On the other hand, stabilizing agents are introduced to control and stabilize the required morphology of metal nanostructures. While, in photochemical synthesis, a variety of light irradiation methods are adopted to synthesize metal nanostructures. Following is the detail of chemical/photochemical synthetic processes in Fig. 5.2 ~ Fig. 5.4.

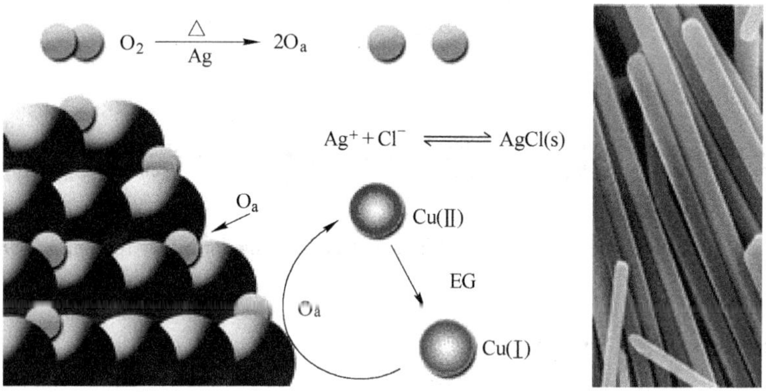

Fig. 5.2 Rapid synthesis of silver nanowires through a CuCl- or $CuCl_2$-mediated polyol process

$$H_2PtCl_6 + 2RCH_2OH \rightarrow Pt^0 + 2RCHO + 6HCl$$

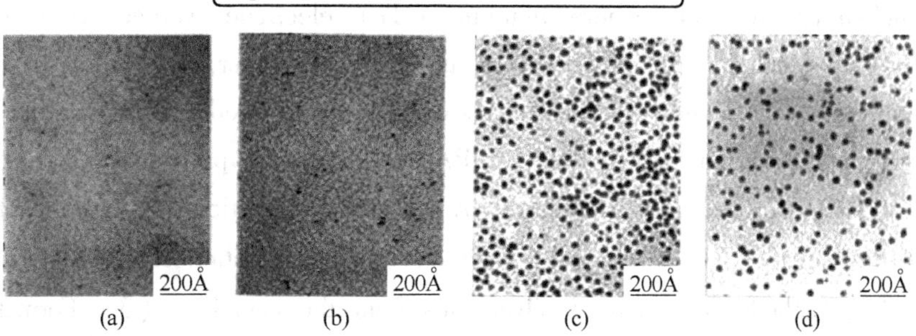

Fig. 5.3 PVP-Pt nanoparticles with different sizes

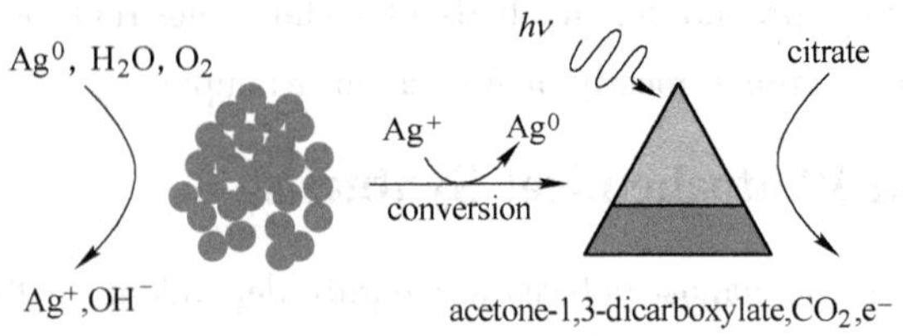

Fig. 5. 4 Room light conversion of citrate stabilized silver nanocrystal seeds to large nanoprisms

5. 1. 1 Light Mediated Synthesis

In this technique, light irradiation is applied to produce nano structures. For example, laser ablation or direct laser irradiation on an aqueous solution of metal salt in the presence of surfactant to fabricate specific shape and size distribution of metal nanoparticles where source of light works as a reducing agent. As shown in Fig. 5. 5, laser light is also used to modify metal nanoparticles by simple melting such as Ag nanospheres into Ag nanoplates, it is known as tailoring process with light. Laser light mediated synthesis is continuing its success for generating highly desirable and well controlled metal nanostructures.

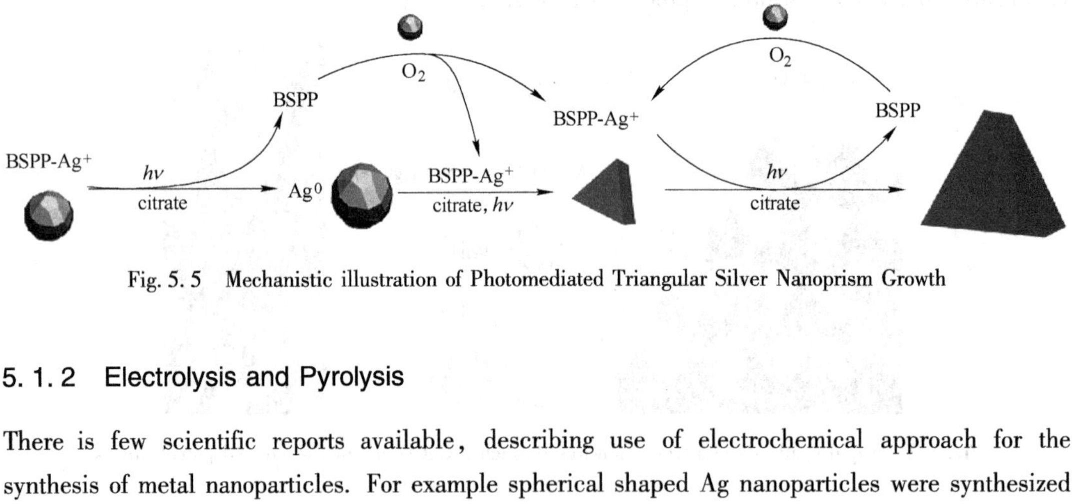

Fig. 5. 5 Mechanistic illustration of Photomediated Triangular Silver Nanoprism Growth

5. 1. 2 Electrolysis and Pyrolysis

There is few scientific reports available, describing use of electrochemical approach for the synthesis of metal nanoparticles. For example spherical shaped Ag nanoparticles were synthesized with an average size of 10nm by reducing Ag^+ ions in the presence of polyvinylpyrrolidone (PVP) using electrochemical method, where titanium (Ti) electrode worked as a cathode and platinum (Pt) plate with 2cm diameter worked as an anode. Silver nanoparticles (up to 10nm) have been synthesized by a method of electrolysis using silver electrodes under conditions of altered current polarity in the sodium polyacrylate (NaPA) solutions. Its proposed multistage scheme of formation of stabilized by polyacrylate anion (PA^-) silver nanoparticles is shown in the Fig. 5. 6, which mainly includes: (1) Ionization of silver and the complexation of produced Ag^+ ions with polyacrylate; (2) Cathode reduction of silver ions from the complex; (3) Formation of PA^- stabilized nanoclusters; (4) Aggregation of nanoclusters and formation of nanoparticles. Ag nanoparticles are also synthesized by adopting another method of spray pyrolysis in which an average grain size of 100nm of Ag nanopowder was synthesized. Among other chemical synthesis, electrolysis and pyrolysis are considered environmental friendly processes as no harmful or toxic reducing agents are used to produce nanostructures.

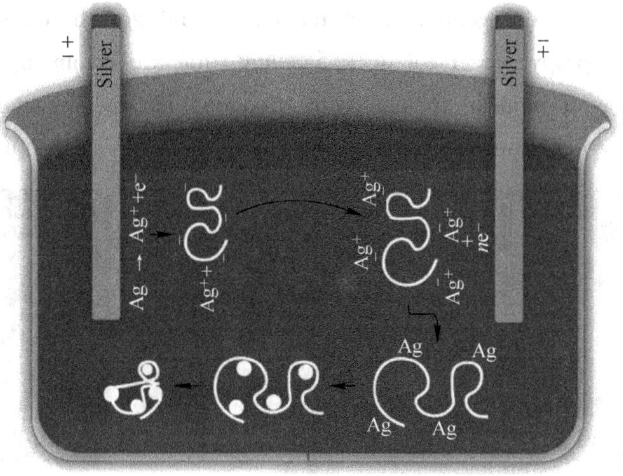

Fig. 5. 6 The Multistage scheme of formation of PA^--stabilized silver nanoparticles

5. 1. 3 Citrate Reduction

One of the popular methods to synthesize Ag colloidal solutions is citrate reduction of Ag^+ ions as we do not need comprehensive laboratory training skills. First time, Lee and Meisel introduced this method in 1982. Generally, Ag nanoparticles are formed, when a calculated amount of sodium citrate is added in anaqueous solution of silver nitrate at boiling and wait for at least 1h. This is a very simple method but did not produce confined size of nanoparticles. The synthesized product shows variety of sizes in the range of 20 ~ 600nm. However, pH control is an important factor, as few shape controlled methods had been discussed by previous researchers by controlling pH values such as at pH = 5. 7 triangular shape and pH = 11. 1 spherical as wellas rod like particles were formed, which is shown in Fig. 5. 7.

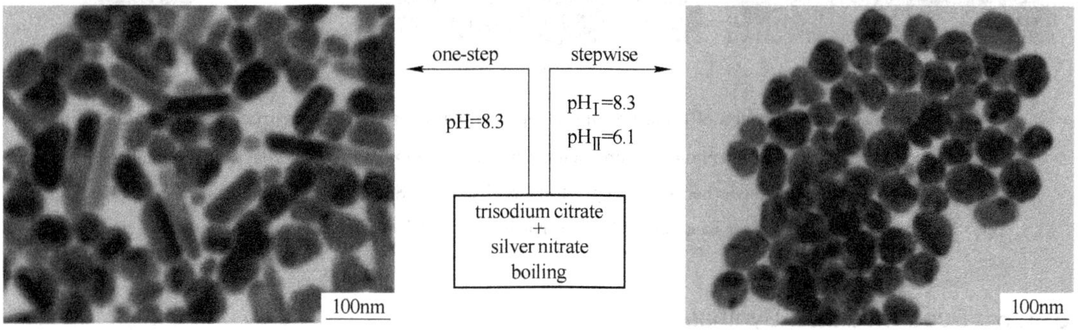

Fig. 5. 7 TEM images of silver nanoparticles synthesized by one-step and stepwise methods under low pH values

5.1.4 Polyol Synthesis

One of the well-known and commonly used methods is to synthesize a wide range of size and shape of Ag nanostructures as shown in Fig. 5.8. In general, Ag salt precursor and capping agents are introduced into polyols for the generation of nucleation and growth towards Ag nanostructure. Propylene glycol, 1,2-propylene glycol or 1,5-pentanediol are commonly used reducing agents which reduce Ag^+ ions present in the solution. Reaction parameters such as temperature and concentrations are very important to overcome the possible control on the final reaction product.

Fig. 5.8 Representative images of electron microscopy of synthesized Ag nanostructures, demonstrating that diverse sizes and morphologies are made possible by controlling the reaction chemistry

(a) Silver nanosphere; (b) Silver necklaces; (c) Silver nanobars; (d) Silver nanocubes; (e) Silver nanoprism; (f) Silver bipyramids; (g) Silver nanostar; (h) Silver nanowire; (i) Silver nanoparticle embedded silica particle

5.1.5 Seed Mediated Synthesis

In recent years, another approach to synthesize metal nanoparticles gaining much attention is seed mediated synthesis in which nanocrystals work as seeds for further growth. This method has advantages on the control over the end product morphology which are shown in Fig. 5.9. For example, Ag nanocubes were synthesized by using spherical or cubic shaped single crystal seed for specific edge length from 30~200nm. A seed-mediated growth approach to improved monodispersity of spheroidal and rod-like gold particles involves controlling nucleation and growth in solution. Monodisperse size-controlled faceted pentagonal silver nanorods were synthesized by thermal regrowth of decahedral silver nanoparticle, and the width of the silver nanorods was determined by the size of the starting decahedral particle. Uniform silver nanowires were synthesized by reducing $AgNO_3$ with ethylene glycol in the presence of seeds and poly (vinyl pyrrolidone), and silver nanowires were with diameters in the range of 30~40nm, and lengths up to ~50μm.

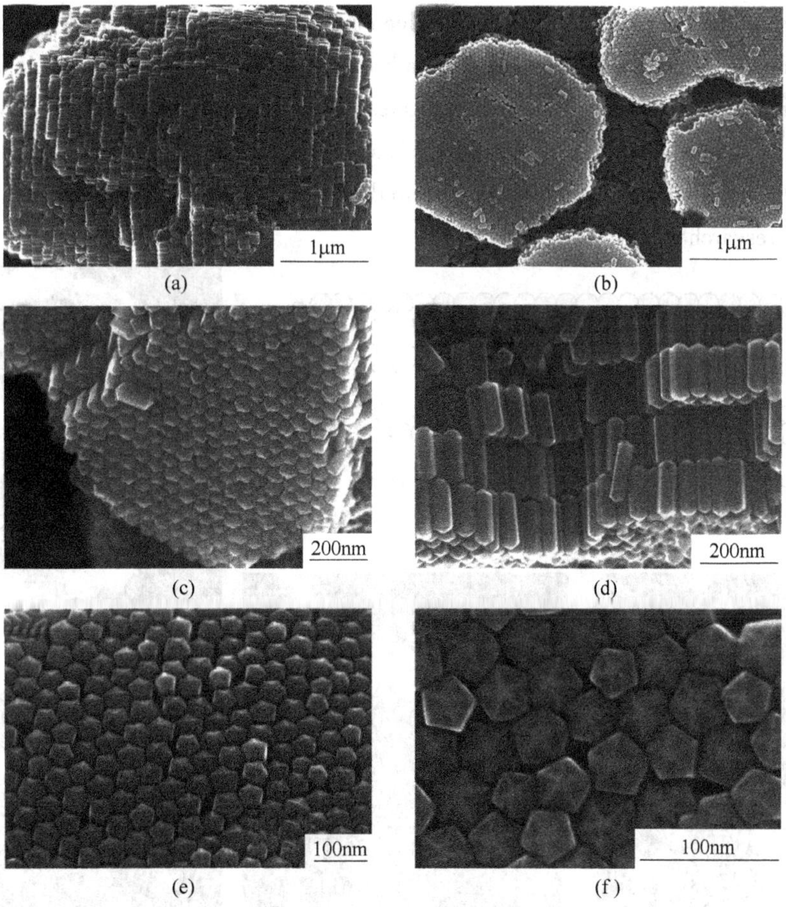

Fig. 5.9 SEM images of self-assembled packing of monodisperse faceted pentagonal rod AgNPs
(a),(c) The rod length is 102nm; (b),(e) The rod length is 123nm; (d),(f) The rod length is 142nm

5.2 Soft and Hard Template Mediated Synthesis

Template based synthesis has been used to prepare specific shape and size of nanostructures which highly depend upon the selected template. There are two types of templates, soft and hard templates. Commonly, surfactant molecules work as a soft templates and form at critical micelle concentration (CMC). Porous anodicaluminum oxide (AAO) membrane could be a good example of hard templates. These micelle and reverse micelles structures present in the solutions are responsible for the shape and size of final product. Previously published research work shows that soft template assisted methods have been involved in preparation of metal nanostructures of various shapes and sizes such as Ag nanowires, nanorods, hollow spheres, nanoplates etc. Based on properties, surfactants can be classified as ionic surfactants, nonionic surfactants, and polymer surfactants. Ionic surfactants include cetyltrimethylammonium bromide (CTAB), octadecyltrimethylammonium chloride (OTAC), disodium (2-ethylhexyl) sulfosuccinate (AOT), and sodium dodecylsulfate (SDS). Nonionic surfactants involve oleic acid (OA), oleylamine (OAM), trioctylphosphine (TOP) and trioctylphosphineoxide (TOPO). Besides, poly(vinyl pyrrolidone) (PVP), poly(vinyl alcohol) (PVA) and poly(ethylene oxide) are also applied. Template assisted synthesis has some advantages such as final product is in well dispersed form due to reduction in the formation of aggregates, reaction conditions are mild as well as benign route in respect of reactants and solvents which is shown in Fig. 5. 10. However, still there is need to improve the parameters to achieve the desired shape and size of nanostructures which is a challenge for researchers.

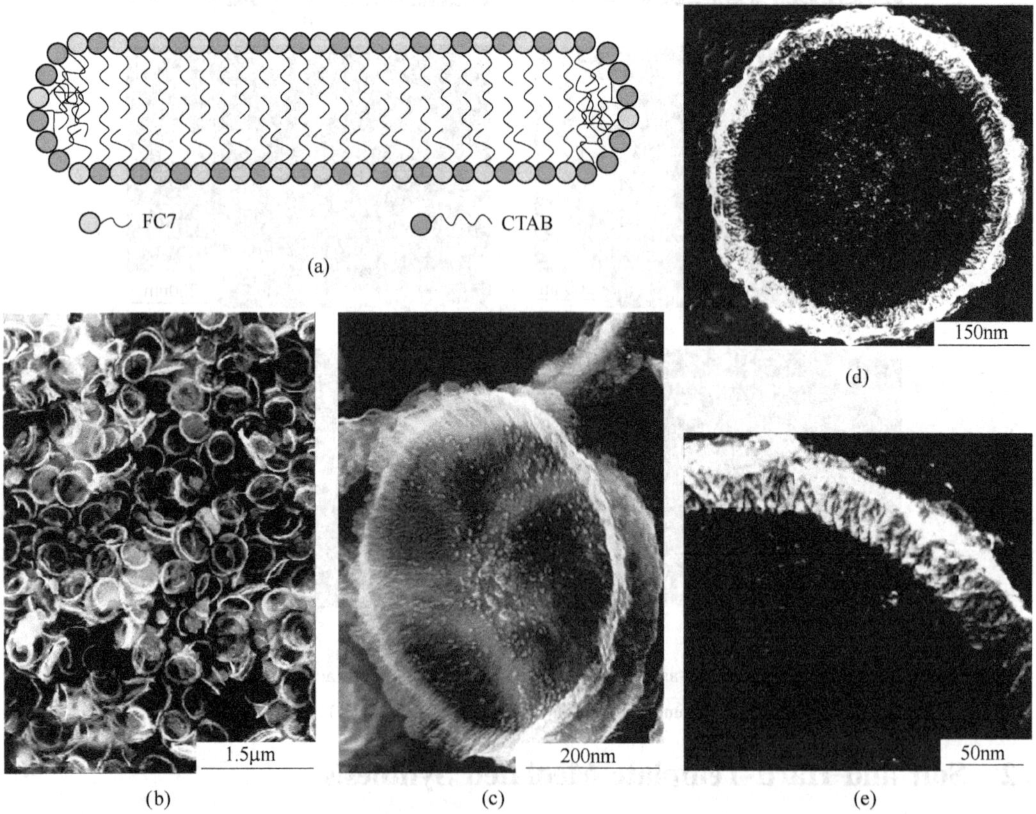

Fig. 5. 10 (a) Illustration of the cross-sectional view of a bicelle composed of two different surfactants, CTAB and sodium perfluorooctanoate in the present system; (b)~(e) SEM images of platinum nanowheels at different magnifications

5.3 Hydrothermal Process

Hydrothermal process is simple and has been widely used for the synthesis of different kinds of nanostructures. Usually, high temperature is required to proceed it. For example, as shown in Fig. 5. 11, Yan et al. synthesized triangular shaped Ru nanoplates with thickness ~ 3nm using $RuCl_3 \cdot H_2O$ and HCHO (w = 40%) in the presence of PVP at 160℃ for 4h. It was also noticed that only change in the concentration of Ru salt and PVP, irregular shape formed with thickness ~ 1.5nm. On the other hand, when Ru salt was replaced with silver nitrate, triangular shaped Ag nanoplates were formed with sharp and curved corners in the presence of PVP at 160℃ for 6h.

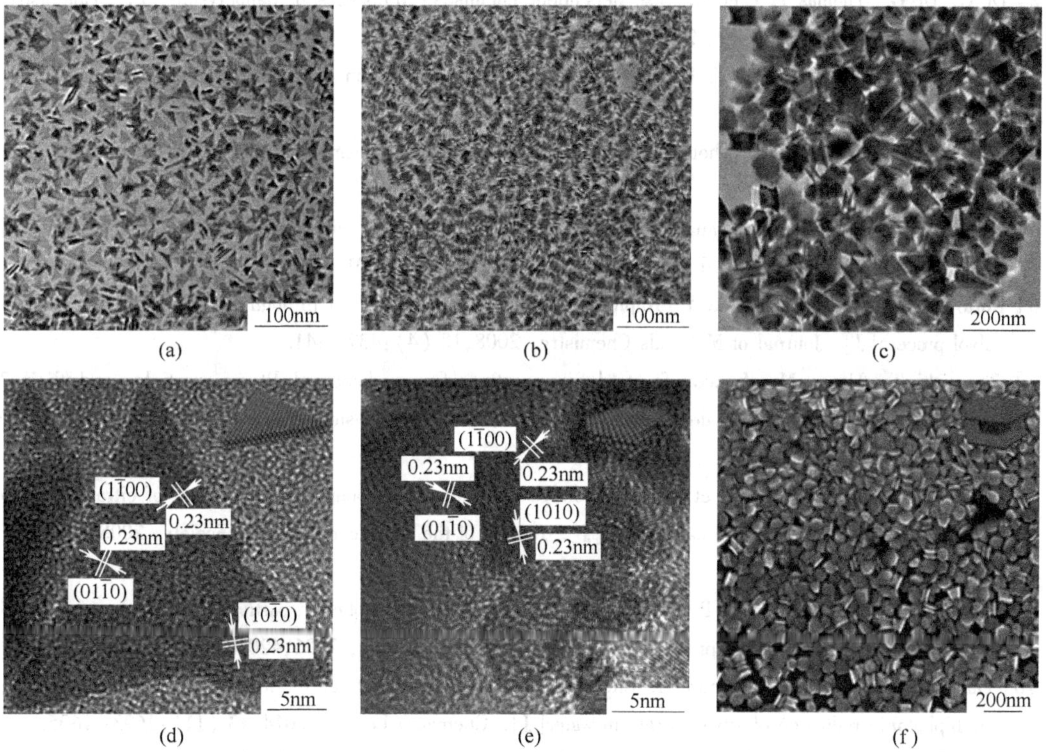

Fig. 5. 11 TEM (a) and HRTEM (b) images of triangle Ru nanoplates, TEM (c) and HRTEM (d) images of ultrathin Ru nanoplates with irregular planer shapes, TEM (e) and SEM (f) images of Ru-capped columns

Recently, it has been noticed that living microorganisms such as bacteria, fungi and plants have great potential for the synthesis of semiconductor(s)/metal(s). The use of microorganisms is environmental friendly and benign synthesis route for which provides good control over size distribution of nanostructures.

Questions

1. What is light mediated synthesis? Please make the appropriate example to illustrate the detailed procedure.
2. Which instrument can be employed to implement electrolysis reduction, preparing the metals?

3. What are commonly used reducing agents when chemical methods are applied to synthesize metals?
4. How to control the microstructure of metals through soft template mediated synthesis? Please take several examples.
5. What is the fundamental principle for the preparation of metals with hydrothermal methods?

References

[1] Munir K, Biesiekierski A, Wen C, et al. In metallic biomaterials processing and medical device manufacturing [M]. Cambridge: Woodhead Publishing, 2020.

[2] Dr C, Dr G, Thomas P J, et al. Size dependent chemistry: properties of nanocrystals [J]. Chemistry-A European Journal, 2002, 8 (1): 28~35.

[3] Gramotnev D K, Bozhevolnyi S I. Plasmonics beyond the diffraction limit [J]. Nature Photonics, 2010, 4 (2): 83~91.

[4] Stewart M E, Anderton C R, Thompson L B, et al. Nanostructured plasmonic sensors [J]. Chemical Reviews, 2008, 108 (2): 494~521.

[5] Tao A, Sinsermsuksakul P, Yang P. Polyhedral silver nanocrystals with distinct scattering signatures [J]. Angewandte Chemie International Edition, 2006, 45 (28): 4597~4601.

[6] Korte K E, Skrabalak S E, Xia Y. Rapid synthesis of silver nanowires through a CuCl- or $CuCl_2$-mediated polyol process [J]. Journal of Materials Chemistry, 2008, 18 (4): 437~441.

[7] Teranishi T, Hosoe M, Tanaka T, et al. Size control of monodispersed Pt nanoparticles and their 2D organization by electrophoretic deposition [J]. The Journal of Physical Chemistry B, 1999, 103 (19): 3818~3827.

[8] Wu X, Redmond P L, Liu H, et al. Photovoltage mechanism for room light conversion of citrate stabilized silver nanocrystal seeds to large nanoprisms [J]. Journal of the American Chemical Society, 2008, 130 (29): 9500~9506.

[9] Lim P Y, Liu R S, She P L, et al. Synthesis of Ag nanospheres particles in ethylene glycol by electrochemical-assisted polyol process [J]. Chemical Physics Letters, 2006, 420 (4): 304~308.

[10] Qazi U Y, Kajimoto S, Fukumura H. Effect of sodium dodecyl sulfate on the formation of silver nanoparticles by biphotonic reduction of silver nitrate in water [J]. Chemistry Letters, 2014, 43 (11): 1693~1695.

[11] Xue C, Métraux G S, Millstone J E, et al. Mechanistic study of photomediated triangular silver nanoprism growth [J]. Journal of the American Chemical Society, 2008, 130 (26): 8337~8344.

[12] Zheng X, Xu W, Corredor C, et al. Laser-induced growth of monodisperse silver nanoparticles with tunable surface plasmon resonance properties and a wavelength self-limiting effect [J]. The Journal of Physical Chemistry C, 2007, 111 (41): 14962~14967.

[13] Jian Z, Li S, Wu J, et al. Plasmon-mediated synthesis of silver triangular bipyramids [J]. Angewandte Chemie International Edition, 2009, 48 (42): 7927~7931.

[14] Sharma V K, Yngard R A, Lin Y. Silver nanoparticles: green synthesis and their antimicrobial activities [J]. Advances in Colloid and Interface Science, 2009, 145 (1): 83~96.

[15] Kuntyi O I, Kytsya A R, Mertsalo I P, et al. Electrochemical synthesis of silver nanoparticles by reversible rurrent in solutions of sodium polyacrylate [J]. Colloid and Polymer Science, 2019, 297 (5): 689~695.

[16] Zhang Q, Liu H, Wang X, et al. Fabrication and characterization of nano silver powder prepared by spray pyrolysis [J]. Journal of Wuhan University of Technology-Mater. 2009, 24 (6): 27~30.

[17] Pillai Z S, Kamat P V. What factors control the size and shape of silver nanoparticles in the citrate ion reduction method? [J]. Journal of Physical Chemistry B, 2004,108 (3):945~951.

[18] Henglein A, Giersig M. Formation of colloidal silver nanoparticles: capping action of citrate[J]. Journal of Physical Chemistry B, 1999,103 (44): 9533~9539.

[19] Dong X, Ji X, Wu H, et al. Shape control of silver nanoparticles by stepwise citrate reduction[J]. The Journal of Physical Chemistry C, 2009,113 (16):6573~6576.

[20] Wiley B, Herricks T, Sun Y, et al. Polyol synthesis of silver nanoparticles: use of chloride and oxygen to promote the formation of single-crystal, truncated cubes and tetrahedrons[J]. Nano Letters, 2004,4 (9): 1733~1739.

[21] Hanisch M, Mačković M, Taccardi N, et al. Synthesis of silver nanoparticle necklaces without explicit addition of reducing or templating agents[J]. Chemical Communications, 2012,48 (36): 4287~4289.

[22] Wiley B J, Chen Y, McLellan J M, et al. Synthesis and optical properties of silver nanobars and nanorice [J]. Nano Letters, 2007,7 (4):1032~1036.

[23] Sun Y, Xia Y. Shape-controlled synthesis of gold and silver nanoparticles[J]. Science, 2010,298 (10): 2176~2179.

[24] Xue B, Wang D, Zuo J, et al. Towards high quality triangular silver nanoprisms: improved synthesis, six-tip based hot spots and ultra-high local surface plasmon resonance sensitivity[J]. Nanoscale, 2015, 7 (17):8048~8057.

[25] Wiley B J, Xiong Y, Li Z Y, et al. Right bipyramids of silver: a new shape derived from single twinned seeds[J]. Nano Letters, 2006,6 (4):765~768.

[26] Tran H N, Nghiem T L, Vu T D, et al. Optical nanoparticles: synthesis and biomedical application[J]. Advances in Natural Sciences Nanoscience & Nanotechnology, 2015,6 (2):023002.

[27] Qiao M, Li J, Chen T, et al. One-dimensional Ag-CoNi nanocomposites modified with amorphous $Sn(OH)_2/SnO_2$ shells for broadband microwave absorption[J]. Journal of Colloid and Interface Science, 2021, 604: 616~623.

[28] Hahm E, Cha M G, Kang E J, et al. Multilayer Ag-embedded silica nanostructure as a surface-enhanced raman Scattering-based chemical sensor with dual-function internal standards[J]. ACS Applied Materials & Interfaces, 2018,10 (47): 40748~40755.

[29] Zhang Q, Li W, Moran C, et al. Seed-mediated synthesis of Ag nanocubes with controllable edge lengths in the range of 30-200nm and comparison of their optical properties[J]. Journal of the American Chemical Society, 2010,132 (32): 11372~11378.

[30] Jana N R, Gearheart L, Murphy C J. Seed-mediated growth approach for shape-controlled synthesis of spheroidal and rod-like gold nanoparticles using a surfactant template[J]. Advanced Materials, 2001, 13 (18): 1389~1393.

[31] Pietrobon B, McEachran M, Kitaev V. synthesis of size-controlled faceted pentagonal silver nanorods with tunable plasmonic properties and self-assembly of these nanorods[J]. ACS Nano, 2009,3 (1):21~26.

[32] Sun Y, Yin Y, Mayers B T, et al. Uniform silver nanowires synthesis by reducing $AgNO_3$ with ethylene glycol in the presence of seeds and poly(vinyl pyrrolidone) [J]. Chemistry of Materials, 2002,14 (11): 4736~4745.

[33] Zhu C, Du D, Eychmüller A, et al. Engineering ordered and nonordered porous noble metal nanostructures: synthesis, assembly, and their applications in electrochemistry[J]. Chemical Reviews, 2015,115 (16): 8896~8943.

[34] Song Y, Dorin R M, Garcia R M, et al. Synthesis of platinum nanowheels using a bicellar template[J].

Journal of the American Chemical Society, 2008,130 (38): 12602~12603.

[35] Darr J A, Zhang J, Makwana N M, et al. Continuous hydrothermal synthesis of inorganic nanoparticles: applications and future directions[J]. Chemical Reviews, 2017,117 (17):11125~11238.

[36] Yin A X, Liu W C, Ke J, et al. Ru nanocrystals with shape-dependent surface-enhanced raman spectra and catalytic properties: controlled synthesis and DFT calculations [J]. Journal of the American Chemical Society, 2012,134 (50):20479~20489.

[37] Kumar A, Kumar V. Biotemplated inorganic nanostructures: supramolecular directed nanosystems of semiconductor(s)/metal(s) mediated by nucleic acids and their properties[J]. Chemical Reviews, 2014, 114 (14):7044~7078.

Chapter 6 Preparation Methods of Inorganic Nonmetallic Material

The interest in the properties of solids and the development of new materials has given rise to the development of a huge variety of methods for preparing them. The method chosen for any solid will depend not only on the composition of the solid but also on the form it is required in for its proposed use. For example, silica glass for fiber optics needs to be much freer of impurities than silica glass used to make laboratory equipment. Some methods may be particularly useful for producing solids in forms that are not the stable form under normal conditions; for example, the synthesis of diamond employs high pressures. Other methods may be chosen because they favor the formation of unusual oxidation states, for example, the preparation of chromium dioxide by the hydrothermal method, or because they promote the production of fine powders or, by contrast, large single crystals. For industrial use, a method that does not employ high temperatures could be favored because of the ensuing energy savings.

6.1 Preparation Methods of Inorganic Powders

6.1.1 Co-precipitation and Precursor Methods

At a simple level, precursors such as nitrates and carbonates can be used as starting materials instead of oxides: they decompose to the oxides on heating at relatively low temperatures, losing gaseous species, and leaving behind fine, more reactive powders. An even more intimate mixture of starting materials can be made by theco-precipitation of solids. A stoichiometric mixture of soluble salts of the metal ions is dissolved and then precipitated as hydroxides, citrates, oxalates, or formats. This mixture is filtered, dried, and then heated to give the final product.

The precursor method achieves mixing at the atomic level by forming a solid compound, the precursor, in which the metals of the desired compound are present in the correct stoichiometry. For example, if an oxide MM'_2O_4 is required, a mixed salt of an oxyacid such as acetate containing M and M′ in the ratio of 1 : 2 is formed. The precursor is then heated to decompose it to the required product. Homogeneous products are formed at relatively low temperatures. A disadvantage is that it is not always possible to find a suitable precursor, but the preparation of barium titanate gives a good illustration of this method.

Barium titanate, $BaTiO_3$, is a ferroelectric material widely used in capacitors because of its high dielectric constant. It was initially prepared by heating barium carbonate and titanium dioxide at high temperature.

$$BaCO_3(s) + TiO_2(s) = BaTiO_3(s) + CO_2(g)$$

However, for modern electronic circuits, it is important to have a product of controlled grain size and the precursor method is one way to achieve this (Another way is the sol-gel method which has also been applied to this material). The precursor used is an oxalate. The first step in the preparation is to prepare an oxo-oxalate of titanium. Excess oxalic acid solution is added to titanium but oxide which initially hydrolyses to give a precipitate which then redissolves in the excess oxalic acid. Barium chloride solution is then added and barium titanyl oxalate precipitates.

$$Ti(OBu)_4(aq) + 4H_2O(l) = Ti(OH)_4(s) + 4BuOH(aq)$$

$$Ti(OH)_4(s) + (COO)_2^{2-}(aq) = TiO(COO)_2(aq) + 2OH^-(aq) + H_2O(l)$$

$$Ba^{2+}(aq) + (COO)_2^{2-}(aq) + TiO(COO)_2(aq) = Ba[TiO((COO)_2)_2](s)$$

This precipitate contains barium and titanium in the correct ratio and is easily decomposed by heat, to give the oxide. The temperature used for this final heating is 920K.

$$Ba[TiO((COO)_2)_2](s) = BaTiO_3(s) + 2CO_2(g) + 2CO(g)$$

The decomposition of oxalates is also used to prepare ferrites MFe_2O_4, which are important as magnetic materials. The products of the precursor method are usually crystalline solids, often containing small particles of large surface area. For some applications, such as catalysis and barium titanate capacitors, this is an advantage. Recently, a novel tartaric acid co-precipitation process was firstly reported to prepare tetragonal $BaTiO_3$ nano-powder with barium chloride, titanium tetrachloride as raw materials and tartaric acid as precipitant agent. The as-obtained $BaTiO_3$ powder exhibited a small average particle size about 140nm, which still held a high tetragonality (c/a ratio) of 1.0092, suggesting its promising applications in multilayer ceramic capacitors.

6.1.2 The Sol-gel Method

The precipitation methods always have the disadvantage that the stoichiometry of the precipitate(s) may not be exact if one or more ions are left in solution. The sol-gel method overcomes this because the reactants never precipitate out. First, a concentrated solution or colloidal suspension of the reactants, the "sol", is prepared, which is then concentrated or matured to form the "gel". This homogeneous gel is then heat-treated to form the product. The main steps in the sol-gel process are outlined in Fig. 6.1.

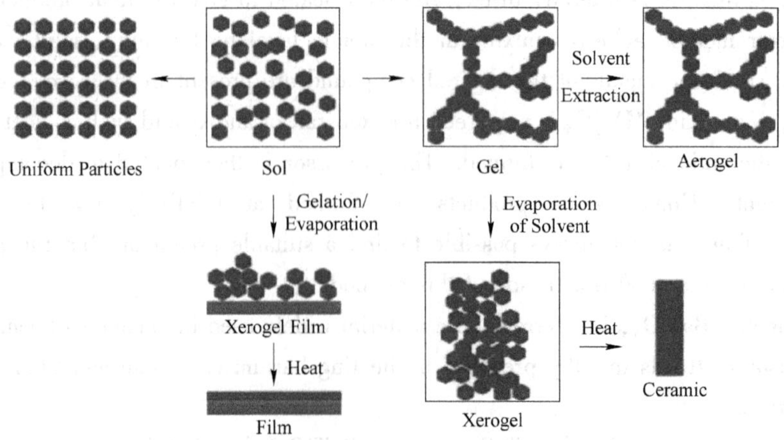

Fig. 6.1 The schematic illustration of sol-gel process

The first investigation of a sol-gel process for synthesis was made in the mid-19th century. This early investigation studied the preparation of silica glass from a sol made by hydrolyzing an alkoxide of silicon. Unfortunately, to prevent the product cracking and forming a fine powder, an aging period of a year or more was required! The sol-gel method was further developed in the 1950s and 1960s after it was realized that colloids, which contain small particles (1 ~ 1000nm in diameter), could be highly chemically homogeneous. A sol is a colloidal suspension of particles in a liquid; for the materials being discussed here, these particles will typically be 1 to 100nm in diameter. A gel is a semi-rigid solid in which the solvent is contained in a framework of material which is either colloidal (essentially a concentrated sol) or polymeric.

To prepare solids using the sol-gel method, a sol of reactants is first prepared in a suitable liquid. Sol preparation can be either simply the dispersal of an insoluble solid or addition of a precursor which reacts with the solvent to form a colloidal product. A typical example of the first is the dispersal of oxides or hydroxides in water with the pH adjusted so that the solid particles remain in suspension rather than precipitate out. A typical example of the second method is the addition of metal alkoxides to water; the alkoxides are hydrolyzed giving the oxide as a colloidal product. The sol is either then treated or simply left to form a gel over time by dehydrating and/or polymerizing. To obtain the final product, the gel is heated. This heating serves several purposes—it removes the solvent, it decomposes anions such as alkoxides or carbonates to give oxides, it allows rearrangement of the structure of the solid, and it allows crystallization to occur. Both the time and the temperature needed for reaction in sol-gel processes can reduced from those in the direct ceramic method; in favorable cases, the time from days to hours, and the temperature by several hundred degrees.

Several examples discussed below illustrate the method; two of the examples have been chosen because they have interesting properties and uses. Many other materials have been prepared by the sol-gel method, and other sol-gel preparations have been employed for the materials chosen. For example, hollow porous SiO_2 microspheres are shown in Fig. 6.2. Therefore, these examples should be taken as illustrative and not as the main uses of the method.

6.1.3 Hydrothermal and Solvothermal Methods

The original hydrothermal method involves heating the reactants in a closed vessel, an autoclave, with water. An autoclave is usually constructed from thick stainless steel to withstand the high pressures, and is fitted with safety valves; it may be lined with non-reactive materials such as Teflon. The autoclave is heated, the pressure increases, and the water remains liquid above its normal boiling temperature of 373K, so-called "super-heated water". These conditions, in which the pressure is raised above atmospheric pressure and the temperature is raised above the boiling temperature of water (but not to as high a temperature as used in the methods described previously), are known as hydrothermal conditions. Hydrothermal conditions exist in nature, and numerous minerals, including naturally occurring zeolites, are formed by this process. Synthetic emeralds are also made under hydrothermal conditions. The term has been extended to other

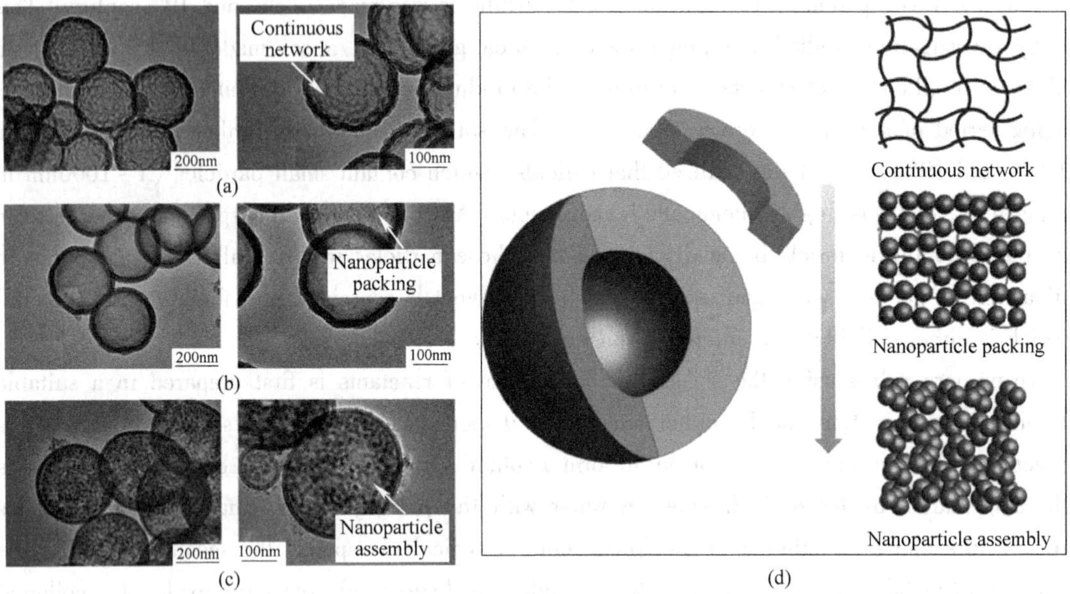

Fig. 6. 2 TEM images (a~c) and schematic illustration of the variation of the shell matrix structure of hollow porous SiO_2 microspheres prepared with sol-gel method (d)

systems with moderately raised pressures and temperatures lower than those typically used in ceramic and sol-gel syntheses. The lower temperatures used are one of the advantages of this method. Others include the preparation of compounds in unusual oxidation states or phases which are stabilized by the raised temperature and pressure. It is useful in metal oxide systems where the oxides are not soluble in water at atmospheric pressure but dissolve in the super-heated water of the hydrothermal set up. Where even these temperatures and pressures are insufficient to dissolve the starting materials, alkali or metal salts can be added whose anions form complexes with the solid and render it soluble. The examples discussed next are chosen to illustrate these advantages and the different variations of the method.

The first industrial process to use hydrothermal methods of synthesis was the production of quartz crystals for use as oscillators in radios. Quartz, SiO_2, can be used to generate a high frequency alternating current via the piezoelectric effect. The hydrothermal growth of quartz crystals employs a temperature gradient. In this variant of hydrothermal processing, the reactant dissolves at a higher temperature, is transported up the reaction tube by convection, and crystallizes out at a lower temperature. A schematic diagram of the apparatus is presented in Fig. 6. 3.

One end of the autoclave is charged with silica and an alkaline solution. When the autoclave is heated, the silica is dissolved in the solution and is transported along the autoclave to the cooler part where it crystallizes out. The alkaline solution returns to the hotter region where it can dissolve more silica. In this case the autoclave, which is made of steel, can act as the reaction vessel. Because of the corrosive nature of superheated solutions and the possibility of contamination of the product with material from the autoclave walls, it is generally necessary to either line the

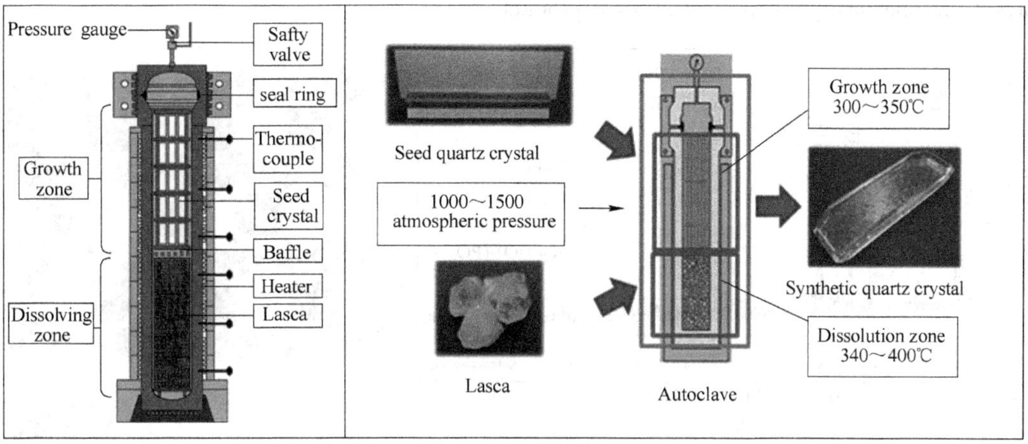

(a) (b)

Fig. 6. 3 Growth of quartz crystals in an autoclave under hydrothermal conditions

autoclave with an inert substance, such as Teflon, or to perform the reaction in a sealed ampoule in the autoclave. Naturally occurring zeolite minerals are formed through a hydrothermal process geochemically, and it was demonstrated in the 1940s and 1950s that they could be synthesized in the laboratory by simulating these hydrothermal conditions. A general method of preparing zeolites involves mixing an alkali, aluminum hydroxide and silica sol, or an alkali, a soluble aluminate and silica sol. The silica and aluminate condense to form a gel as in the sol-gel method but the gel is then heated in a closed vessel at temperatures close to 373K. Under these conditions, zeolites instead of other alumina-silicate phases crystallize out. For example, in the synthesis of a typical zeolite, zeolite A ($Na_{12}[(AlO_2)_{12}(SiO_2)_{12}] \cdot 27H_2O$) hydrated alumina, $Al_2O_3 \cdot 3H_2O$, is dissolved in concentrated sodium hydroxide solution. The cooled solution is then mixed with a solution of sodium metasilicate, $Na_2SiO_3 \cdot 9H_2O$, and a thick white gel form. The gel is preparative methods 163 then placed in a closed Teflon bottle and heated to about 363K over 6 hours. The reaction time can be reduced by using microwave radiation for this final step: for example, zeolite A has been made by heating for 45 minutes using pulses of 300 watt-power microwave radiation .

Hydrothermal recrystallization is not confined to quartz. Some ⅡB-ⅥA compounds, such as ZnSe, ZnTe, CdSe, and CdTe, have also been recrystallized into large single crystals using the technique, as well as aluminum orthophosphate, $AlPO_4$. The technique can also be used in the case of preparation of potassium titanyl phosphate. Fig. 6. 4 shows XRD patterns and schematic diagrams of the monoclinic structure.

$$KH_2PO_4 + TiO_2 \longrightarrow KTiOPO_4 + H_2O$$

Hydrothermal synthesis does not require the water to be above its critical point. Huan et al. published a synthesis of $VOC_6H_5PO_3 \cdot H_2O$ prepared from phenyl phosphonic acid, $C_6H_5PO(OH)_2$, and vanadium(Ⅲ) oxide, V_2O_3. The two reagents were added to water, sealed in a Teflon acid digestion bomb, and heated to 200℃. Pure water has a vapor pressure of 225 PSI at 200℃, well below the bursting pressure of the bomb (1800 PSI). Unlike the quartz example, in this case, the solvent became incorporated into the final product.

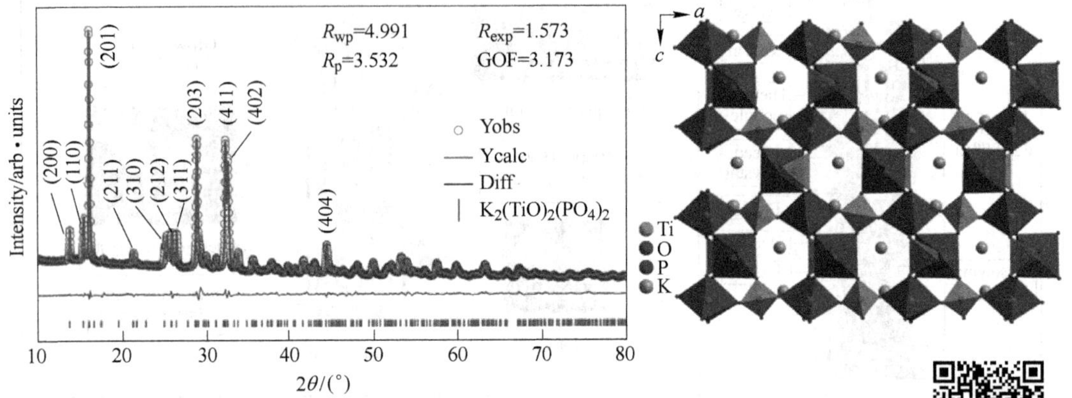

(a) (b)

Fig. 6. 4 Structure characterizations of the as-synthesized $KTiOPO_4$ compound

(a) Rietveld-refined XRD profiles, experimental (dark cyan dots), fitted (orange line), residual discrepancy (gray lower line) and Bragg peaks (light red line); (b) Schematic diagrams of the monoclinic structure

Scan the QR code to see color picture

The solvothermal method uses solvents above their boiling points, but not necessarily above their critical points. The method can be employed using sealed glass tubes, acid digestion bombs, or high-pressure autoclaves. A popular method is to use Teflon digestion bombs and microwave heaters. As in the hydrothermal method, the solvent can become part of the final product. Li et al. have reviewed the use of the solvothermal method for preparing metal chalcogenides in an ethylenediamine solvent. Ethylenediamine has a normal boiling point of 117℃ and a critical temperature and pressure of 320℃ and 913 PSI, respectively. A typical reaction (unbalanced) is described in the following.

$$CuCl_2 + Sb_2Se_3 + C_2N_2H_8 \longrightarrow Cu_2SbSe_3(C_2N_2H_8)$$

where $C_2N_2H_8$ is ethylenediamine, $H_2NCH_2CH_2NH_2$. The reaction temperature would be less than 250℃, the maximum operating temperature for the Teflon-lined digestion bomb. This compound, $Cu_2SbSe_3(C_2N_2H_8)$, is an interesting example of a broad class of compounds, known as intercalation compounds. Besides, monodisperse water-dispersible porous magnetic Fe_3O_4 microspheres were prepared via a simple hydrothermal method together with a gas-bubble-assisted Ostwald ripening process, and their morphology was shown in the Fig. 6. 5 (a), (b). Moreover, the hydrothermal reaction of $KMnO_4$ was applied to grow the flower-like MnO_2 shells on the surface of Fe_3O_4 microspheres, forming core-shell Fe_3O_4@MnO_2 microspheres in the Fig. 6. 5 (c), (d), which exhibit superior microwave absorption performance.

6. 1. 4 Molten Salts

Salts, both those with high melting points and those with low melting points, have several different uses in the preparation of inorganic materials. The term molten salt is generally taken to mean an inorganic salt with a high melting point relative to room temperature. Molten salts may be used in

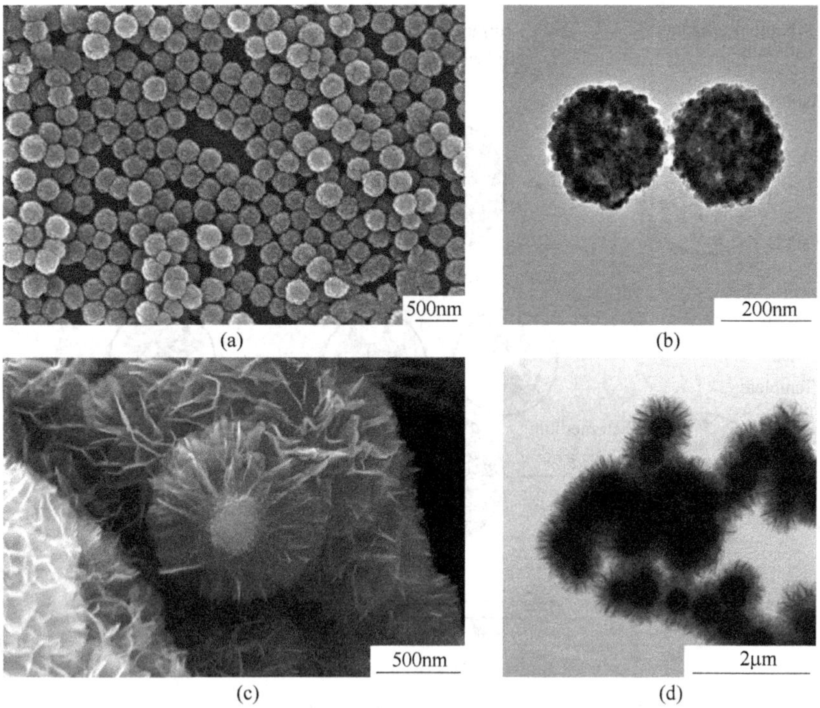

Fig. 6.5 SEM(a) and TEM(b) images of porous Fe_3O_4 microspheres; SEM(c) and TEM(d) images of Fe_3O_4@MnO_2 microspheres

topochemical and nontopochemical reactions of existing solids and their mechanisms are shown in Fig. 6.6. Topochemical reactions (sometimes called topotactic reactions) follow the pathway of minimum atomic or molecular movement. Consequently, the products are isostructural, or nearly so, with the starting solid phase. In a topochemical synthesis, a molten salt is reactive (it serves as a reactant) because a component of it is incorporated into the final product. An example of a topochemical route utilizing a reactive molten salt is ion exchange in pure alkali metal halides and nitrates. In this method, a solid phase, which is usually in powder form because of the large surface area/volume ratio per particle, with easily exchangeable cations (those with layered and tunnel structures) is immersed in the molten salt. The salt functions as a source of mobile cations that interchange with those of the solid, to an extent determined by the equilibrium constant for the reaction. The rate of ion exchange is determined largely by the ion mobilities, which is a function of temperature and ion charge (higher charged ions are less mobile), while thermodynamics usually favor exchange of ions of similar size and/or higher charge. It is possible to carry out aliovalention exchange, whereby cations of different charges are exchanged (e.g. an alkalimetal cation and an alkaline-earth metal cation) with the concurrent creation of vacancies (v) to maintain overall electroneutrality in the final product. For example:

$$Na_2La_2Ti_3O_{10}(s) + Ca(NO_3)_2(l) \longrightarrow CaLa_2Ti_3O_{10}(s) + 2NaNO_3(l)$$

$$K = \frac{\alpha_{Na}^2 + \alpha_{Ca}}{\alpha_{Na} + \alpha_{Ca}}$$

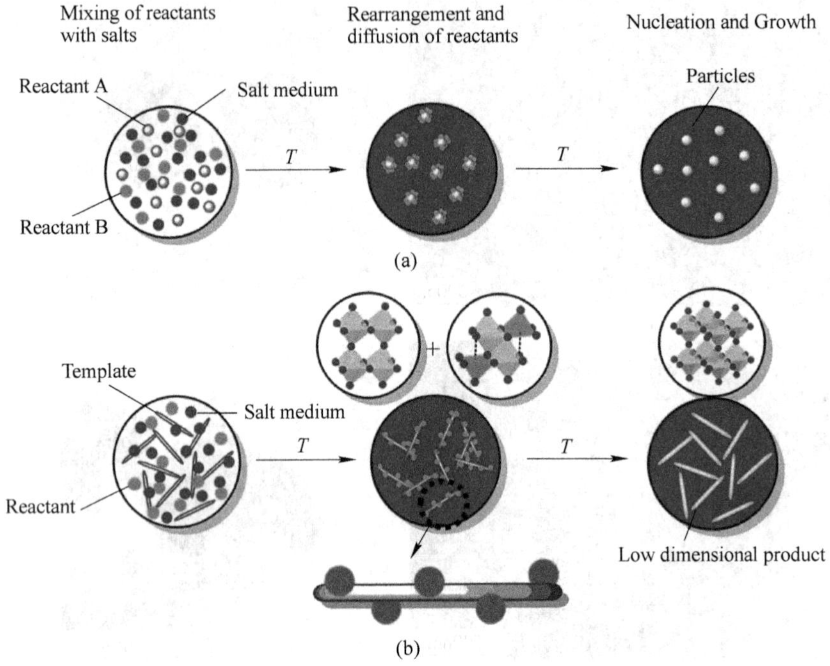

Scan the QR code to see color picture

Fig. 6. 6 Schematic showing the mechanisms

(a) The conventional molten salts synthesis; (b) Topochemical molten salts synthesis routes

In this equilibrium constant expression, the activity coefficient, a, is specified for the exchangeable cations (Na^+, Ca^{2+}) in each phase. The equilibrium may lie to the left or right, depending on the value of K. Accordingly, the product is more accurately represented as $Na_{2-x}Ca_x/2vx/2La_2Ti_3O_{10}$, to signify possible incomplete, or partial, ion exchange. The creation of vacancies in a host by aliovalent ion exchange opens the material to the possibility of subsequent intercalation reactions, where another species can be inserted into the location once occupied by the original cation. In some cases, the intercalation can introduce mixed valency on transition metal cations that are present in the structure. This will depend on the inserted species' reduction potential relative to those of the transition metal in its various oxidation states. It is noteworthy that many of these products are kinetically stable, thermodynamically metastable phases that cannot be prepared by the direct high-temperature ceramic route.

A momentary digression is warranted here concerning the phase purity of powder products. What one hopes to obtain in a process such as the ion-exchange reaction discussed above is a homogeneous single-phase solid solution, as opposed to a multi-phase mixture of product(s) and parent phase? The definition of phase, unfortunately, is somewhat ambiguous, as it gives no criterion for the length scale or degree of homogeneity. Does homogeneous require that every unit cell in every crystallite be identical? Generally speaking, we accept a "phase" as a state of matter in which the physical structure is, on average, microscopically homogeneous and, as such, one that has uniform thermodynamic properties. Phase purity is typically ascertained by conventional X-ray powder diffractometry. A powder diffraction pattern contains peaks whose positions depend on the size and shape of the unitcell, while their intensities are dependent on the identities of the unit cell contents. For coarse-grained powders (crystallite size >1μm), small differences in peak positions, intensities, and asymmetry are easily discernible, which makes the technique well suited for checking phase purity. However, the possible presence of isostructural phases with only slight differences in unit cell dimensions and/or with isoelectronic species (and thus very similar atomic form factors or scattering factors) does present challenges regarding verifying phase purity.

Both reactive and nonreactive molten salts can be used in nontopochemical routes. An example of a nontopochemical route to inorganic materials utilizin greactive molten salts is when a metallic element is reduced in a low-meltingalkali metal polychalcogenide (A_2Q_n, where Q = O, S, Se, Te) to form a ternary metal chalcogenide. Potassium bismuth sulfide (KBi_3S_5) has been prepared in this way from the reaction of bismuth metal in potassium sulfide. Superconductive magnesium boride films (MgB_2) have been prepared by electrolysis from molten mixtures of $MgCl_2$, KCl, and MgB_2O_4, as has CaB_6 from $CaCl_2$ and CaB_4O_7. $BaBiO_3$-derived super conducting oxides have been synthesized from anodic oxidation of molten salt mixtures containing $Ba(OH)_2$, KOH, and Bi_2O_3. In nonreactive molten salts, on the other hand, flux components are not incorporated into the product phase. Here, the molten salt acts more in the classical sense as a reagent to promote the reaction at a lower temperature than would be required by the ceramic, or direct, route. This is accomplished by two attributes of molten salts: an acid-base equilibrium that enables the general dissolution-recrystallization of metal oxides; and a highly electropositive (oxidizing) environment that stabilizes the highest oxidation state of many transition metals, which can lead to mixed valency. A plethora of complex transition metal oxides have been synthesized in nonreactive molten alkali metal hydroxides, carbonates, and hypochlorites. Examples of such molten salt routes to mixed transition metal oxides include:

(1) La_2CuO_4: from a mixture of La_2O_3 and CuO in molten NaOH;

(2) $YBa_2Cu_4O_8$: from Y_2O_3, BaO, and CuO in Na_2CO_3/K_2CO_3 flux under flowing oxygen;

(3) $La_4Ni_3O_{10}$: created by bubbling chlorine gas through a NaOH solution of lanthanum and nickel nitrates.

6. 1. 5 Combustion Synthesis

Combustion synthesis in solid-state chemistry is also known as metathesis reactions or self-propagating high-temperature synthesis (SHS). Our emphasis in the ceramic method was the need for high reaction temperatures in order to achieve usable diffusion coefficients for the atoms in the reacting solids. Mechanical alloying offers an alternative to high reaction temperatures (i. e. , high-pressure pulses), which themselves could be a source for local rapid heating. In the combustion synthesis, high temperatures again take center stage, but this time the high temperature is achieved not by placing the sample in a furnace, but rather, by using the heat generated by the exothermic chemical reaction itself.

Combustion synthesis can be operated in two modes, viz. propagation and volume reaction (also called thermal explosion). In the propagation mode, the reaction is initiated by ignition at one point and then spreads over the sample by the propagation of reaction front (Fig. 6.7). In the volume reaction mode, the whole sample is heated to the ignition temperature so that combustion reaction occurs simultaneously everywhere in the sample. In practice, it is difficult to ensure that the temperature is strictly uniform and the reaction happens exactly at the same time throughout the sample, and volume reaction is a propagation process with a very high propagation velocity.

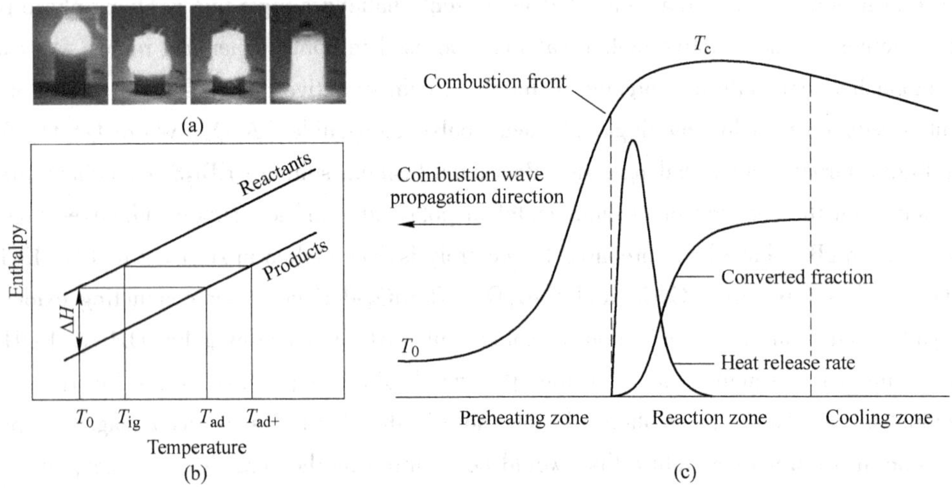

Fig. 6.7 (a) Photographs showing the propagation of reaction front in combustion synthesis; (b) A schematic illustration of enthalpy-temperature correlation in combustion reactions; (c) A typical structure in the region neighboring the reaction front

In a trivial application of this method, one could consider the synthesis of carbon dioxide from methane as a combustion synthesis. This trivial example provides some framework for additional insights into the solid-state case. The balanced reaction for methane reacting with oxygen is

$$CH_4(g) + 2O_2(g) \longrightarrow CO_2(g) + 2H_2O(g)$$

This reaction could be viewed in terms of a double displacement reaction: One of the oxygen atoms and one of the hydrogen atoms have displaced each other. One of the most famous examples of this type of exchange reaction, a single exchange, occurs between iron oxide and aluminum:

$$Fe_2O_3(s) + 2Al(s) \longrightarrow 2Fe(s) + Al_2O_3(s)$$

This reaction, known as the thermite reaction, requires an initial source of heat to begin the reaction, but after ignition, this highly exothermic reaction generates enough heat to proceed in a self-sustained manner. The iron is produced as a liquid because of the intense heat. After enough liquid iron has been produced, the bottom is opened and liquid iron brows out. In this case, the iron is being used to weld two railroad rails together.

The topic combustion synthesis was first studied by Alexander Merzhavov in the 1970s has been reviewed by Gillan and Kaner 1996. A typical reaction occurs between a metal halide and an alkali or alkaline earth main group compound, such as an oxide, nitride, or sulfide. The reaction of $ZrCl_4$ and lithium nitride is such an example:

$$3ZrCl_4(s) + 4Li_3N(s) \longrightarrow 3ZrN(s) + 12LiCl(s)$$

Zirconium chloride is a white powder, and lithium nitride is a black powder. When this reaction proceeds, the lithium chloride forms as a liquid just as their on forms as a liquid in the thermite reaction. The melting point of lithium chloride is 605℃ (the melting point of iron is 1536℃). The temperature of the zirconium chloride/lithium nitride reaction reaches 1370℃ less than 1 second after the reaction is initiated. This liquid plays the same role as the flux agents already discussed: It provides a medium for rapid mixing of reagents.

Alkali or alkaline earth metals, although common in combustion syntheses, are not required for a combustion synthesis. $TiCl_3$ and $AlCl_3$ can be combined in a combustion reaction to form Ti_3Al, $TiAl_3$ or TiAl, depending on the initial molar amounts of the reactants. Ferrite magnetic materials of barium ferrite ($BaFe_{12}O_{19}$), lithium ferrite ($Li_{0.5}Fe_{2.5}O_4$) and magnesium zinc ferrite ($Mg_{0.5}Zn_{0.5}Fe_2O_4$) have been prepared by self-propagating high-temperature synthesis (SHS) reactions from iron, iron(Ⅲ) oxide and metal oxides and peroxides:

$$BaO_2 + 6Fe + 3Fe_2O_3 + 4O_2 \longrightarrow BaFe_{12}O_{19}$$

$$MgO + ZnO + 2Fe + Fe_2O_3 + NaClO_4 \longrightarrow 2Mg_{0.5}Zn_{0.5}Fe_2O_4(s) + NaCl + 1/2O_2(g)$$

The use of metal nitrates provides a particularly convenient method for doping a host oxide with luminescent ions. Along with the yttrium nitrate, a luminescent rare earth ion such as Eu^{3+}, can be added as a nitrate salt. In addition to yttria, alumina can also be prepared from aluminum nitrate and hydrazine:

$$4Al(NO_3)_3(aq) + 15N_2H_4(aq) \longrightarrow 2Al_2O_3(s) + 30H_2O(g) + 21N_2(g)$$

Combustion synthesis uses highly exothermic ($\Delta H < -170$kJ/mol) and even explosive reactions to maintain a self-propagating high reaction temperature and has been used to prepare many refractory materials including borides, nitrides, oxides, silicides, intermetallics, and ceramics. The reactants are mixed, formed into a pellet, and then ignited (laser, electric arc, heating coil) at high temperature. Once ignited the reaction propagates as a wave, the synthesis wave, through the pellet, and the reaction must lose less heat than it generates or it will quench; temperatures up to 3000K are maintained during the fast reactions. Fig. 6.8 depicts the synthesis wave moving through a mixture of MgO, Fe, Fe_2O_3, and $NaClO_4$, the images were captured by a thermal imaging camera. Self-ignition can sometimes be achieved by ball milling. The method is now used commercially in Russia, Spain, and Korea because it is fast, economical, and gives high purity products. Thermal ignition of a mixture of BaO_2, Fe_2O_3, Cr_2O_3 and Fe produces a ferrite $BaFe_{12-x}Cr_xO_{19}$; in the presence of a magnetic field the reaction proceeds faster and with higher temperature propagation wave. Examples of other useful products of these reactions are:

(1) Hydrides (e.g., MgH_2) for hydrogen storage;

(2) Borides (e.g., TiB_2) for abrasives and cutting tools;

(3) Carbides (e.g., SiC and TiC) for abrasives and cutting tools;

(4) Nitrides (e.g., Si_3N_4) for high strength, heat resistant ceramics;

(5) Oxides (e.g., cuprates) for high temperature superconductors;

(6) Silicides (e.g., $MoSi_2$) for high temperature heating elements;

(7) Composites (e.g., the thermite reaction, $3Fe_3O_4+8Al = 4Al_2O_3+9Fe$) have been used to

coat the inside of steel pipes with an inner layer of Fe, and a surface layer of Al_2O_3, using a centrifugal process.

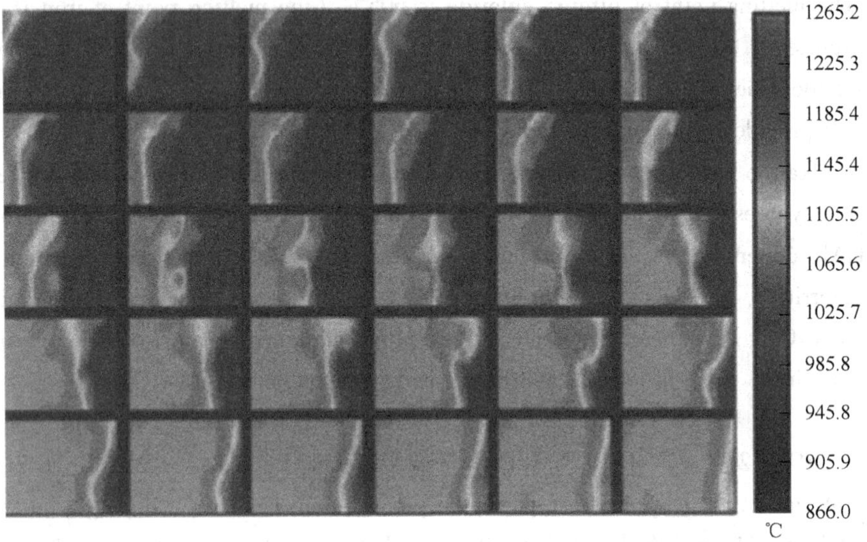

Fig. 6. 8 Thermal images of the synthesis wave moving through a pellet of MgO, Fe, Fe_2O_3, and $NaClO_4$

(Each image is of dimension 3×2mm; Images were captured at 0. 06s intervals; The first image is top left and the last is bottoming right)

6. 2 Molding Process of Inorganic Materials

6. 2. 1 Fabrication and Processing of Glasses

Before we discuss specific glass-forming techniques, some of the temperature-sensitive properties of glass materials must be presented. Glassy, or noncrystalline, materials do not solidify in the same sense as do those that are crystalline. Upon cooling, a glass becomes more and more viscous in a continuous manner with decreasing temperature; there is no definite temperature at which the liquid transforms into a solid as with crystalline materials. In fact, one of the distinctions between crystalline and noncrystalline materials lies in the dependence of specific volume (or volume per unit mass, the reciprocal of density) on temperature, as illustrated in Fig. 6. 9(a); this same behavior is exhibited by highly crystalline and amorphous polymers. For crystalline materials, there is a discontinuous decrease in volume at the melting temperature T_m. However, for glassy materials, volume decreases continuously with temperature reduction; a slight decrease in slope of the curve occurs at what is called the glass transition temperature, or fictivetemperature, T_g. Below this temperature, the material is a glass; above it, the material is first a supercooled liquid and, finally, a liquid.

Also important in glass-forming operations are the viscosity-temperature characteristics of the glass. Fig. 6. 9(b) plots the logarithm of viscosity versus the temperature for fused silica, high silica, borosilicate, and soda-lime glasses. On the viscosity scale, several specific points are

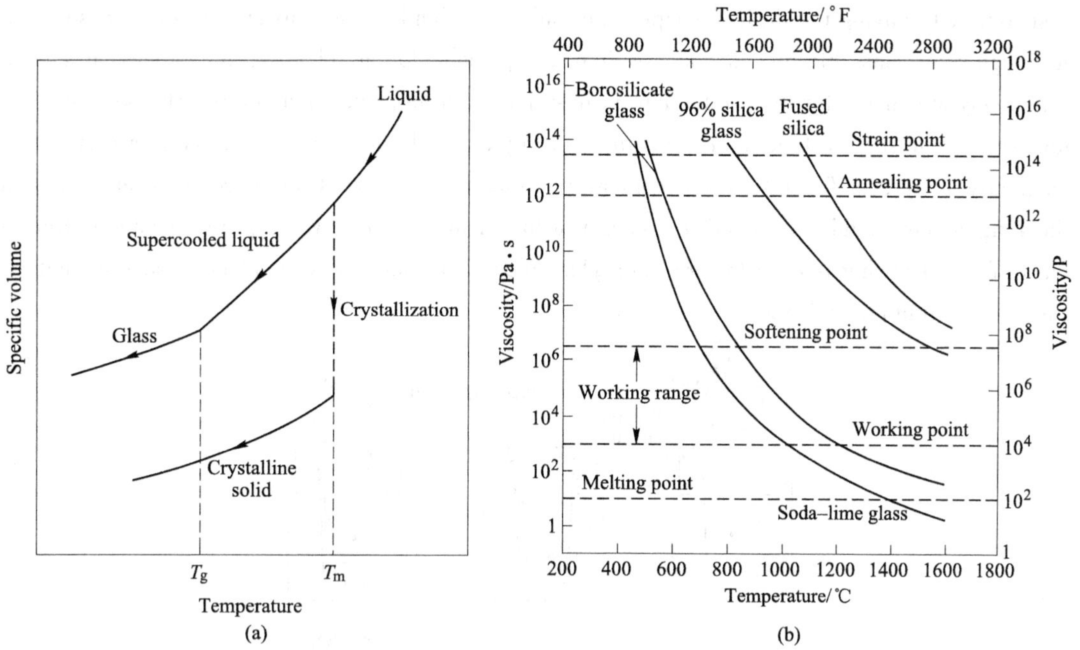

Fig. 6.9 (a) Contrast of specific volume-versus-temperature behavior of crystalline and noncrystalline materials (Crystalline materials solidify at the melting temperature T_m; Characteristic of the noncrystalline state is the glass transition temperature T_g); (b) Logarithm of viscosity versus temperature for fused silica and three silica glasses

important in the fabrication and processing of glasses, including melting point, working point, softening point, annealing point and strain point. Most glass-forming operations are carried out within the working range—between the working and softening temperatures. The temperature at which each of these points occurs depends on glass composition. For example, from Fig. 6.9(b), the softening points for soda-lime and 96% silica glasses are about 700℃ and 1550℃ (1300℉ and 2825℉), respectively. That is, forming operations may be carried out at significantly lower temperatures for the soda-lime glass. The formability of a glass is tailored to a large degree by its composition.

Glass is produced by heating the raw materials to an elevated temperature above which melting occurs. Most commercial glasses are of the silica-soda-lime variety; the silica is usually supplied as common quartz sand, whereas Na_2O and CaO are added as soda ash (Na_2CO_3) and limestone ($CaCO_3$). For most applications, especially when optical transparency is important, it is essential that the glass product be homogeneous and pores free. Homogeneity is achieved by complete melting and mixing of the raw ingredients. Porosity results from small gas bubbles that are produced; these must be absorbed into the melt or otherwise eliminated, which requires proper adjustment of the viscosity of the molten material.

Five different forming methods are used to fabricate glass products: pressing, blowing, drawing, and sheet and fiber forming. Pressing is used in the fabrication of relatively thick-walled pieces such as plates and dishes. The glass piece is formed by pressure application in a graphite-coated cast iron mold having the desired shape; the mold is typically heated to ensure an even surface. Although some glass blowing is done by hand, especially for art objects, the process has been completely automated for the production of glass jars, bottles, and light bulbs. The several steps involved in one such technique are illustrated in Fig. 6. 10. From a raw gob of glass, a parison, or temporary shape, is formed by mechanical pressing in a mold. This piece is inserted into a finishing or blow mold and forced to conform to the mold contours by the pressure created from a blast of air. Drawing is used to form long glass pieces that have a constant cross section such as sheet, rod, tubing, and fibers.

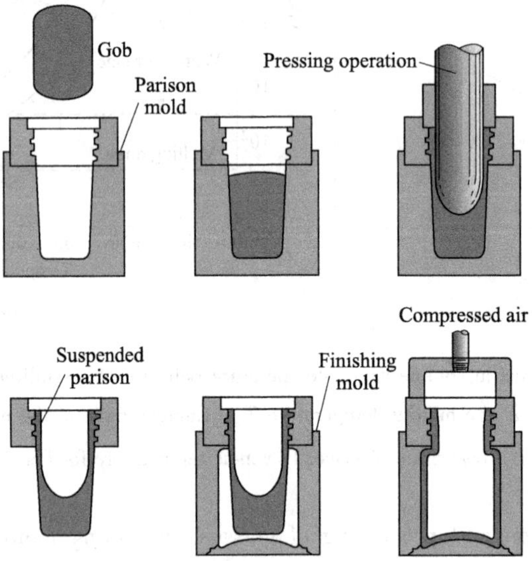

Fig. 6. 10 The press-and-blow technique for producing a glass bottle

Until the late 1950s, sheet glass (or plate) was produced by casting (or drawing) the glass into a plate shape, grinding both faces to make them flat and parallel, and, finally, polishing the faces to make the sheet transparent—a procedure that was relatively expensive. A more economical float process was patented in 1959 in England. With this technique (represented schematically in Fig. 6. 11), the molten glass passes (on rollers) from one furnace onto a bath of liquid tin located in a second furnace. Thus, as this continuous glass ribbon "floats" on the surface of the molten tin, gravitational and surface tension forces cause the faces to become perfectly flat and parallel and the resulting sheet to be of uniform thickness. Furthermore, sheet faces acquire a bright, "fire-polished" finish in one region of the furnace. The sheet next passes into an annealing furnace (lehr), and is finally cut into sections (Fig. 6. 11). The success of this operation requires rigid control of both temperature and chemistry of the gaseous atmosphere. Continuous glass fibers are formed in a rather sophisticated drawing operation. The molten glass is contained in a platinum heating chamber. Fibers are formed by drawing the molten glass through many small orifices at the chamber base. The glass viscosity, which is critical, is controlled by chamber and orifice temperatures.

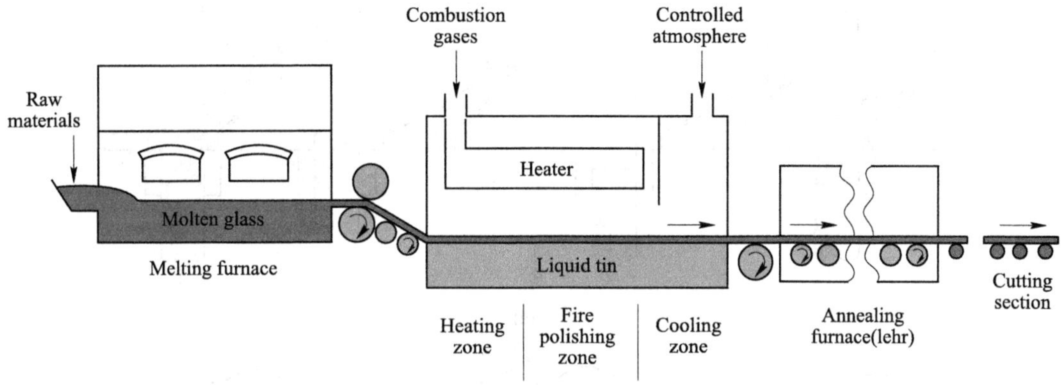

Fig. 6.11 Schematic diagram showing the float process for making sheet glass

6.2.2 Powder Processing

Several ceramic-forming techniques have already been discussed relative to the fabrication of glass and clay products. Another important and commonly used method that warrants brief treatment is powder pressing. Powder pressing—the ceramic analogue to powder metallurgy—is used to fabricate both clay and nonclay compositions, including electronic and magnetic ceramics, as well as some refractory brick products. In essence, a powdered mass, usually containing a small amount of water or other binder, is compacted into the desired shape by pressure. The degree of compaction is maximized and the fraction of void space is minimized by using coarse and fine particles mixed in appropriate proportions. There is no plastic deformation of the particles during compaction; as there may be with metal powders. One function of the binder is to lubricate the powder particles as they move past one another in the compaction process.

There are three basic powder-pressing procedures: uniaxial, isostatic (or hydrostatic), and hot pressing. For uniaxial pressing, the powder is compacted in a metal die by pressure that is applied in a single direction. The formed piece takes on the configuration of the die and platens through which the pressure is applied. This method is confined to shapes that are relatively simple; however, production rates are high and the process is inexpensive. The steps involved in one technique are illustrated in Fig. 6.12(a).

For isostatic pressing, the powdered material is contained in a rubber envelope and the pressure is applied isostatically by a fluid (i.e., it has the same magnitude in all directions). More-complicated shapes are possible than with uniaxial pressing; however, the isostatic technique is more time consuming and expensive.

For both uniaxial and isostatic procedures, a firing operation is required after the pressing operation. During firing the formed piece shrinks and experiences a reduction of porosity and an improvement in mechanical integrity. These changes occur by the coalescence of the powder particles into a denser mass in a process termed sintering. The mechanism of sintering is schematically illustrated in Fig. 6.12(b). After pressing, many of the powder particles touch one

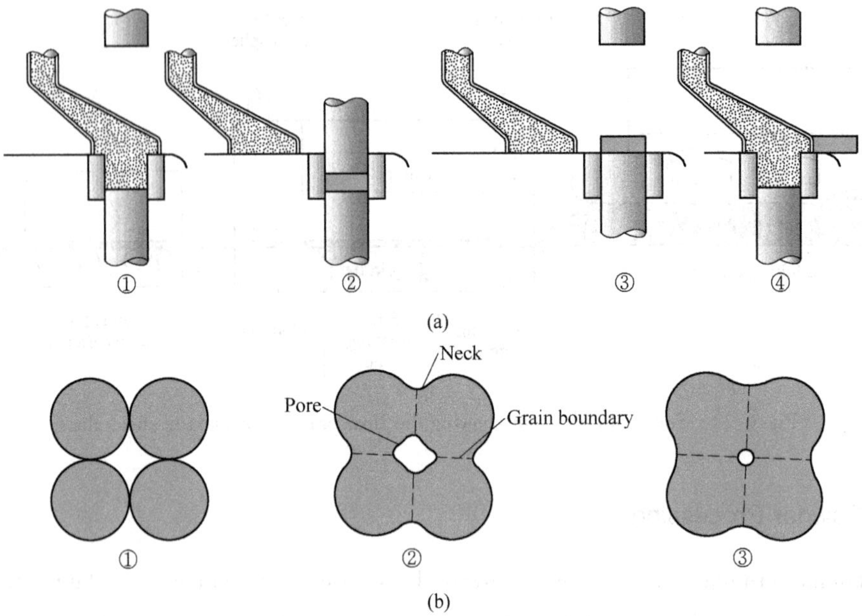

Fig. 6. 12 (a) Schematic representation of the steps in uniaxial powder pressing; (b) For a powder compact, microstructural changes that occur during firing

another. During the initial sintering stage, necks form along the contact regions between adjacent particles; in addition, a grain boundary forms within each neck and every interstice between particles becomes a pore. As sintering progresses, the pores become smaller and more spherical. A scanning electron micrograph of a sintered alumina material is shown in Fig. 6. 13(a). The driving force for sintering is the reduction in total particle surface area; surface energies are larger in magnitude than grain boundary energies. Sintering is carried out below the melting temperature, so that a liquidphase is normally not present. The mass transport that is necessary to affect the changes shown in Fig. 6. 13(b) is accomplished by atomic diffusion from the bulk particles to the neck regions.

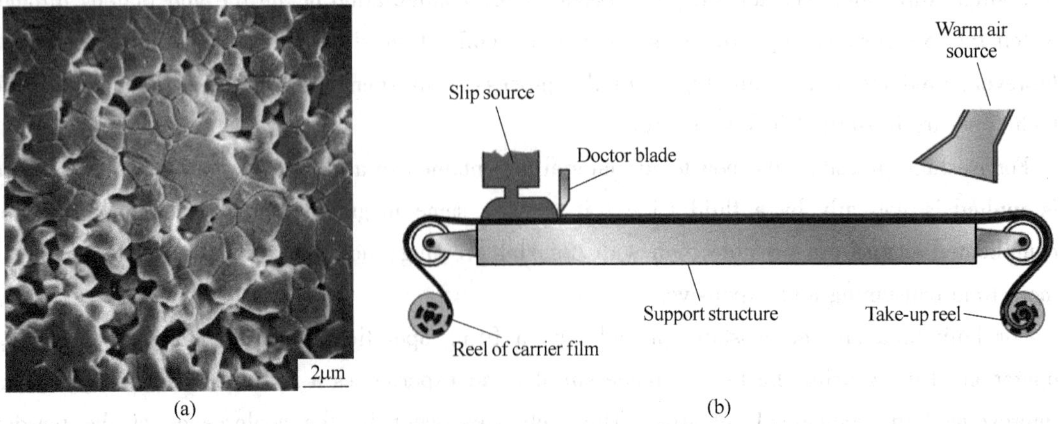

Fig. 6. 13 (a) Scanning electron micrograph of an aluminum oxide powder compact that was sintered at 1700℃ for 6min; (b) Schematic diagram showing the tape-casting process using a doctor blade

With hot pressing, the powder pressing and heat treatment are performed simultaneously—the powder aggregate is compacted at an elevated temperature. The procedure is used for materials that do not form a liquid phase except at very high and impractical temperatures; in addition, it is used when high densities without appreciable grain growth are desired. This is an expensive fabrication technique and has some limitations. It is costly in terms of time because both mold and die must be heated and cooled during each cycle. In addition, the mold is usually expensive to fabricate and typically has a short lifetime.

6.2.3 Tape Casting

Tape casting is an important ceramic fabrication technique. As the name implies, in this technique, thin sheets of a flexible tape are produced by means of a casting process. These sheets are prepared from slips in many respects similar to those employed for slipcasting. This type of slip consists of a suspension of ceramic particles in anorganic liquid that also contains binders and plasticizers, which are incorporated to impart strength and flexibility to the cast tape. De-airing in a vacuum may also be necessary to remove any entrapped air or solvent vapor bubbles, which may act as crack-initiation sites in the finished piece. The actual tape is formed by pouring the slip onto a flat surface (of stainless steel, glass, a polymeric film, or paper); a doctor blade spreads the slip into a thin tape of uniform thickness, as shown schematically in Fig. 6.13(b). In the drying process, volatile slip components are removed by evaporation; this green product is a flexible tape that may be cut or into which holes may be punched prior to a firing operation. Tape thicknesses normally range between 0.1 and 2mm (0.004 and 0.08in). Tape casting is widely used in the production of ceramic substrates that are used for integrated circuits and for multilayered capacitors.

Cementation is also considered a ceramic fabrication process. The cement material, when mixed with water, forms a paste that, after being fashioned into a desired shape, subsequently hardens as a result of complex chemical reactions.

Questions

1. Which synthetic methods would be suitable for producing the following characteristics? a) A thin film of material. b) A single crystal. c) A single crystal containing layers of different material with the same crystal structure. d) A powder of homogeneous composition.
2. What are the advantages and disadvantages of using the sol-gel method to prepare barium titanate for use in a capacitor?
3. In hydrothermal processes involving alumina (Al_2O_3), such as the synthesis of zeolites, alkali is added to the reaction mixture. Suggest a reason for this addition.
4. How does the thickness of a glassware affect the magnitude of the thermal stresses that may be introduced? Why?
5. What is the distinction between glass transition temperature and melting temperature?
6. In your own words, briefly describe what happens as a glass piece is thermally tempered.

7. Some ceramic materials are fabricated by hot isostatic pressing. Cite some of the limitations and difficulties associated with this technique.

References

[1] Hu M Z C, Miller G A, Payzant E A, et al. Homogeneous (co) precipitation of inorganic salts for synthesis of monodispersed barium titanate particles[J]. Journal of Materials Science, 2000,35:2927~2936.

[2] Zhang X, Yue J, Zhao Y, et al. Synthesis of tetragonal $BaTiO_3$ nano-particle via a novel tartaric acid co-precipitation process[J]. Ceramics International, 2021,47 (5): 7263~7267.

[3] Choi H, Stathatos E, Dionysiou D D. Sol-gel preparation of mesoporous photocatalytic TiO_2 films and TiO_2/Al_2O_3 composite membranes for environmental applications[J]. Applied Catalysis B: Environmental, 2006, 63 (1):60~67.

[4] Anastasescu C, Anastasescu M, Teodorescu V S, et al. SiO_2 nanospheres and tubes obtained by sol-gel method[J]. Journal of Non-Crystalline Solids, 2010,356 (44):2634~2640.

[5] Ellerby L, Nishida C, Nishida F, et al. Encapsulation of proteins in transparent porous silicate glasses prepared by the sol-gel method[J]. Science, 1992,255 (5048):1113~1115.

[6] Xing Z, Tay S W, Ng Y H, et al. Porous SiO_2 hollow spheres as a solar reflective pigment for coatings[J]. ACS Applied Materials & Interfaces, 2017,9 (17): 15103~15113.

[7] Walker A C. Hydrothermal synthesis of quartz crystals[J]. Journal of the American Ceramic Society, 1953, 36 (8):250~256.

[8] Bertone J F, Cizeron J, Wahi R K, et al. Hydrothermal synthesis of quartz nanocrystals[J]. Nano Letters, 2003,3 (5): 655~659.

[9] Ansari M, Aroujalian A, Raisi A, et al. Preparation and characterization of nano-NaX zeolite by microwave assisted hydrothermal method[J]. Advanced Powder Technology, 2014,25 (2): 722~727.

[10] Nasser G A, Muraza O, Nishitoba T, et al. Microwave-assisted hydrothermal synthesis of CHA zeolite for methanol-to-olefins reaction[J]. Industrial & Engineering Chemistry Research, 2019,58 (1): 60~68.

[11] Li D, Yao J, Wang H. Hydrothermal synthesis of $AlPO_4$-5: effect of precursor gel preparation on the morphology of crystals[J]. Progress in Natural Science: Materials International, 2012,22 (6): 684~692.

[12] Qi Y, Li J, Zhong W, et al. $KTiOPO_4$: a long-life, high-rate and low-temperature-workable host for Na/K-ion batteries[J]. Chemical Engineering Journal, 2021,417:128159.

[13] Huan G, Johnson J W, Jacobson A J, et al. Hydrothermal synthesis and single-crystal structural characterization of $VO(VO_3)_6(VO(C_{10}H_8N_2)_2)_2$[J]. Journal of Solid State Chemistry, 1991,91 (2): 385~389.

[14] Li J, Chen Z, Wang R J, et al. Low temperature route towards new materials: solvothermal synthesis of metal chalcogenides in ethylenediamine[J]. Coordination Chemistry Reviews, 1999,190~192: 707~735.

[15] Qiao M, Lei X, Ma Y, et al. Facile synthesis and enhanced electromagnetic microwave absorption performance for porous core-shell Fe_3O_4@MnO_2 composite microspheres with lightweight feature[J]. Journal of Alloys and Compounds, 2017, 693: 432~439.

[16] Qiao M, Lei X, Ma Y, et al. Dependency of tunable microwave absorption performance on morphology-controlled hierarchical shells for core-shell Fe_3O_4@MnO_2 composite microspheres[J]. Chemical Engineering Journal, 2016, 304: 552~562.

[17] McCarthy T J, Tanzer T A, Kanatzidis M G. A new metastable three-dimensional bismuth sulfide with large tunnels: synthesis, structural characterization, ion-exchange properties, and reactivity of KBi_3S_5[J].

Journal of the American Chemical Society, 1995,117 (4): 1294~1301.

[18] Abe H, Yoshii K. Electrochemical synthesis of superconductive boride MgB_2 from molten salts[J]. Japanese Journal of Applied Physics, 2002,41 (6): 685~687.

[19] Liu S F, Fu W T. Synthesis of superconducting $Ba_{1-x}K_xBiO_3$ by a modified molten salt process[J]. Materials Research Bulletin, 2001,36 (7/8): 1505~1512.

[20] Gupta S K, Mao Y. A review on molten salt synthesis of metal oxide nanomaterials: status, opportunity, and challenge[J]. Progress in Materials Science, 2021,117:100734.

[21] Liu G, Chen K, Li J. Combustion synthesis: an effective tool for preparing inorganic materials[J]. Scripta Materialia, 2018,157:167~173.

[22] Sun Y, Yao B, He Q, et al. Synthesis and formation mechanism of cubic ZrN nanopowders by mechanochemical reaction of $ZrCl_4$ and Li_3N[J]. Journal of Alloys and Compounds, 2009,479 (1): 599~602.

[23] Blair R G, Gillan E G, Nguyen N K B, et al. Rapid solid-state synthesis of titanium aluminides[J]. Chemistry of Materials, 2003,15 (17): 3286~3293.

[24] Parkin I P, Pankhurst Q A, Affleck L, et al. Self-propagating high temperature synthesis of BaFeO, MgZnFeO and LiFeO; time resolved X-ray diffraction studies (TRXRD) [J]. Journal of Materials Chemistry, 2001,11 (1): 193~199.

[25] Ozuna O, Hirata G A, Mckittrick J. Pressure influenced combustion synthesis of Al_2O_3 nanocrystalline powders[J]. Journal of Physics Condensed Matter, 2004,16 (15):2585~2591.

[26] Patil K C, Aruna S T, Mimani T. Combustion synthesis: an update[J]. Current Opinion in Solid State and Materials Science, 2002,6 (6): 507~512.

[27] Chick L A, Pederson L R, Maupin G D, et al. Glycine-nitrate combustion synthesis of oxide ceramic powders[J]. Materials Letters, 1990,10 (1): 6~12.

[28] Moore J J, Feng H J. Combustion synthesis of advanced materials: part Ⅰ. reaction parameters[J]. Progress in Materials Science, 1995,39 (4): 243~273.

[29] Ye T, Guiwen Z, Weiping Z, et al. Combustion synthesis and photoluminescence of nanocrystalline Y_2O_3: Eu phosphors[J]. Materials Research Bulletin, 1997,32 (5):501~506.

[30] Shand E B. Engineering glass[J]. Modern Materials, 1968,6:247~317.

[31] Phillips C J. Glass, the miracle maker[J]. Journal of Chemical Education, 1941, 19, 398.

Chapter 7 Preparation Mechanisms of Polymer

The formation of synthetic polymers is a process which occurs via chemical connection of many hundreds up to many thousands of monomer molecules. As a result, macromolecular chains are formed. They are, in general, linear, but can be branched, hyperbranched, or crosslinked as well. However, depending on the number of different monomers and how they are connected, homo- or one of the various kinds of copolymers can result. The chemical process of chain formation may be subdivided roughly into two classes, depending on whether it proceeds as a chain-growth or as a step-growth reaction.

7.1 Radical Chain Polymerization

Polymerization of unsaturated monomers by chain polymerization will be discussed in this section. Chain polymerization is initiated by reactive species R^* produced from some compound I termed an initiator:

$$I \longrightarrow R^* \tag{7.1}$$

The reactive species, which may be a free radical, cation, or anion, adds to a monomer molecule by opening the p-bond to form a new radical, cation, or anion center, as the case may be. The process is repeated as many more monomer molecules are successively added to continuously propagate the reactive center:

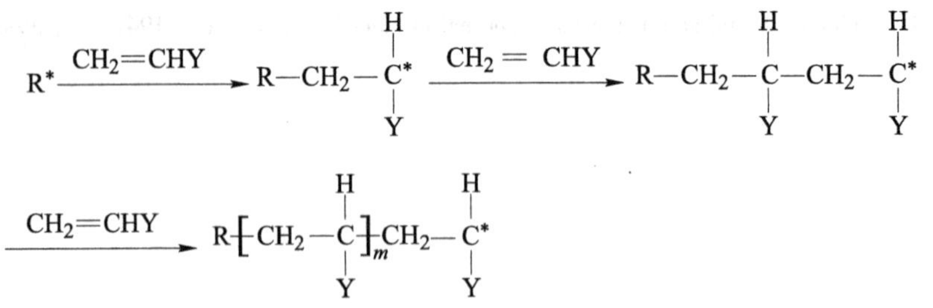

(7.2)

Polymer growth is terminated at some point by destruction of the reactive center by an appropriate reaction depending on the type of reactive center and the particular reaction condition.

7.1.1 Sequence of Radical Chain Polymerization

Radical chain polymerization is chain reaction consisting of a sequence of three steps-initiation, propagation, and termination. The initiation step is considered to involve two reactions. The first is the production of three radicals by any one of a number of reactions. The usual case is the homolytic dissociation of an initiator species I to yield a pair of radicals R· .

$$I \xrightarrow{k_d} 2R\cdot \qquad (7.3)$$

where k_d is the rate constant for the catalyst dissociation. The second part of the initiation involves the addition of this radical to the first monomer molecule to produce the chain initiating radical M_1.

$$R\cdot + M \xrightarrow{k_i} M_1 \qquad (7.4)$$

where M represents a monomer molecule and k_i is the rate constant for the initiation step (7.4). For the polymerization of $CH_2{=}CHY$, Eq. (7.4) takes the form

$$R\cdot + CH_2{=}CHY \longrightarrow R{-}CH_2{-}C(H)(Y)\cdot \qquad (7.5)$$

The radical R· is often referred to as an initiator radical or primary radical to distinguish it from the chain-initiating species ($M_1\cdot$).

Propagation consists of the growth of $M_1\cdot$ by the successive additions of large numbers of monomer molecules according to Eq. (7.5). Each addition creates a new radical that has the same identity as the one previously, except that it is larger by one monomer unit. The successive additions may be represented by

$$\left.\begin{aligned} M_1\cdot + M &\xrightarrow{k_p} M_2\cdot \\ M_2\cdot + M &\xrightarrow{k_p} M_3\cdot \\ M_3\cdot + M &\xrightarrow{k_p} M_4\cdot \\ &\vdots \end{aligned}\right\} \qquad (7.6)$$

where k_p is the rate constant for propagation. Propagation with growth of the chain to high polymer proportions takes place very rapidly. The value of k_p for most monomers is in the range $10^2 \sim 10^4$ L/(mol · s). This is a large rate constant—much larger than those usually encountered in step polymerizations.

At some point, the propagation polymer chain stops growing and terminates. Termination with the annihilation of the radical centers occurs by biomolecular reaction between radicals. Two radicals react with each other by combination (coupling) or, more rarely, by

$$\sim\!\!\sim CH_2{-}C(H)(Y)\cdot + \cdot C(H)(Y){-}CH_2\sim\!\!\sim \xrightarrow{k_{tc}} \sim\!\!\sim CH_2{-}C(H)(Y){-}C(H)(Y){-}CH_2\sim\!\!\sim \qquad (7.7)$$

Disproportionation, in which hydrogen radical which is beta to one radical center, is transferred to another radical center. This results in the formation of two polymer molecules—one saturated and one unsaturated:

$$\sim\!\!\sim CH_2{-}C(H)(Y)\cdot + \cdot C(H)(Y){-}C(H)(Y)\sim\!\!\sim \xrightarrow{k_{td}} \sim\!\!\sim CH_2{-}CH(Y) + C(H)(Y){=}C(Y)\sim\!\!\sim \qquad (7.8)$$

Termination can also occur by a combination of coupling and disproportionation. The two different modes of termination can be represented in general terms by

$$\mathrm{M}_n\cdot + \mathrm{M}_m\cdot \xrightarrow{k_{\mathrm{tc}}} \mathrm{M}_{n+m}\mathrm{M}_n\cdot + \mathrm{M}_m\cdot \xrightarrow{k_{\mathrm{tc}}} \mathrm{M}_n + \mathrm{M}_m \tag{7.9}$$

where k_{tc} and k_{td} are the rate constants for termination by coupling and disproportionation, respectively. One can also express the termination step by

$$\mathrm{M}_n\cdot + \mathrm{M}_m\cdot \xrightarrow{k_{\mathrm{t}}} \text{dead polymer} \tag{7.10}$$

where the particular mode of termination is not specified and

$$k_{\mathrm{t}} = \alpha k_{\mathrm{tc}} + (1-\alpha)k_{\mathrm{td}} \tag{7.11}$$

where α and $(1-\alpha)$ are the fractions of termination by coupling and disproportionation, respectively. The term dead polymer signifies the cessation of growth for the propagation radical. The propagation reaction would proceed indefinitely until all the monomers in a reaction system were exhausted if it were not for the strong tendency toward termination. Typical termination rate constants are in the range of $10^6 \sim 10^8$ L/(mol · s) or orders of magnitude greater than the propagation rate constants. The much greater value of k_{t}(whether k_{tc} or k_{td}) compared to k_{p} does not prevent propagation because the radical species are present in very low concentrations and because the polymerization rate is dependent on only the one-half power of k_{t}.

7.1.2 Kinetics of Initiation and Polymerization

Eqs. (7.3) through (7.11) constitute the detailed mechanism of a free-radical-initiated chain polymerization. The chain nature of the process resides in the propagation step [Eq. (7.6)] in which large numbers of monomer molecules are converted to polymer for each initiating radical produced in the first step[Eqs. (7.4)and(7.5)]. In order to obtain a kinetic expression for the rate of polymerization, it is necessary to assume that k_{p} and k_{t} are independent of the size of the radical. This assumption is inherent in the notation used in Eqs. (7.6) through (7.10). It is exactly the same assumption—the equal-reactivity assumption—that was employed in deriving the kinetics of step polymerization. Very small radicals are more reactive than propagation radicals, but this effect is not important because the effect of size vanishes at the dimer or trimer size. There are limitations to the equal-reactivity assumption with respect to the sizes of propagating radicals.

Monomer disappears by the initiation reaction [Eqs. (7.4) and (7.5)] as well as by the propagation reactions [Eq. (7.6)]. The rate of monomer disappearance, which is synonymous with the rate of polymerization, is given by

$$R_{\mathrm{i}} + R_{\mathrm{p}} = -\frac{\mathrm{d}[\mathrm{M}]}{\mathrm{d}t} \tag{7.12}$$

where R_{i} and R_{p} are the rates of initiation and propagation, respectively. However, the number of monomer molecules reacting in the initiation step is far less than the number in the propagation step ofr a process producing high polymer. To a very close approximation the former can be neglected and the polymerization rate is given simply by the rate of propagation

$$R_{\mathrm{p}} = -\frac{\mathrm{d}[\mathrm{M}]}{\mathrm{d}t} \tag{7.13}$$

The rate of propagation, and therefore the rate of polymerization, is the sum of many individual propagation steps. Since the rate constants for all the propagation steps are the same, one can express the polymerization rate by

$$R_p = k_p[M\cdot][M] \tag{7.14}$$

where [M] is the monomer concentration and [M·] is the monomer concentration and [M·] is the total concentration of all chain radicals, that is, all radicals of size M· and larger.

Eq. (7.14) for the polymerization rate is not directly usable because it contains a term for concentration of radicals. Radical concentrations are difficult to measure quantitatively, since they are very low ($\sim 10^{-8}$ mol/L), and it is therefore desirable to eliminate [M·] from Eq. (7.14). In order to do this, the steady-state assumption is made that the concentration of radicals increases initially, but almost instantaneously reaches a constant, steady-state value. The rate of change of the concentration of radicals quickly becomes and remains zero during the course of the polymerization. This is equivalent to stating that the rates of initiation R_i and termination R_t of radicals are equal or

$$R_i = R_t = 2k_t[M\cdot]^2 \tag{7.15}$$

The steady-state assumption is not unique to polymerization kinetics. It is often used in developing the kinetics of many small-molecule reactions that involve highly reactive intermediates present at very low concentrations-conditions that are present in radical chain polymerizations. The theoretical validity of the steady-state assumption has been discussed and its experimental validity shown in many polymerizations. Typical polymerizations achieve a steady-state after a period, which may be at most a minute.

The right side of Eq. (7.15) represents the rate of termination. There is no specification as to whether termination is by coupling or disproportionation since both follow the same kinetic expression. The use of the factor 2 in the termination rate equation follows the generally accepted convention for reactions destroying radicals in pairs. It is also generally used for reactions creating radicals in pairs as in Eq. (7.3). This convention is now the IUPAC preferred convention. In using the polymer literature, one should be aware that the factor of 2 has not always been followed. Rearrangement of Eq. (7.15) to

$$[M\cdot] = \left(\frac{R_i}{2k_t}\right)^{1/2} \tag{7.16}$$

and substitution into Eq. (7.14) yields

$$R_p = k_p[M]\left(\frac{R_i}{2k_t}\right)^{1/2} \tag{7.17}$$

for the rate of polymerization. It is seen that Eq. (7.17) has the significant conclusion of the dependence of the polymerization rate on the square root of the initiation rate. Doubling the rate of initiation does not double the polymerization rate; the polymerization rate is increased only by the factor $\sqrt{2}$. This behavior is a consequence of the bimolecular termination reaction between radicals.

The rate of producing primary radicals by thermal homolysis of an initiator R_d is given by

$$R_d = 2fk_d[\mathrm{I}] \tag{7.18}$$

where [I] is the concentration of the initiator and f is the initiator efficiency. The initiator efficiency is defined as the fraction of the radicals produced in the homolysis reaction that initiate polymer chains. The value of f is usually less than unity due to wastage reactions. In most polymerizations, the second step (the addition of the primary radical to monomer) is much faster than the first step. The homolysis of the initiator is the rate-determining step in the initiation sequence, and the rate of initiation in then given by

$$R_i = 2fk_d[\mathrm{I}] \tag{7.19}$$

The rate of polymerization is given by Eq. (7.14).

$$R_p = k_p[\mathrm{M}]\left(\frac{fk_d[\mathrm{I}]}{2k_t}\right)^{1/2} \tag{7.20}$$

7.2 Step-growth Polymerization

A number of different chemical reactions may be used to synthetic polymeric materials by step polymerization. These include esterification, amidation, the formation of urethanes, aromatic substitution, and others. Polymerization usually proceeds by the reactions between two different functional groups, for example, hydroxyl and carboxyl groups, or isocyanate and hydroxyl groups. Many of the common condensation polymers are listed in Table 7.1.

Table 7.1 Typical condensation polymers

Type	Characteristic linkage	Polymerization reaction
Polyamide	—NH—CO—	$\mathrm{H_2N—R—NH_2 + HO_2C—R'—CO_2H \rightarrow H{+}NH—R—NHCO—R'—CO{+}_nOH + H_2O}$
		$\mathrm{H_2N—R—NH_2 + ClCO—R'—ClCO \rightarrow H—NH—R—NHCO—R'—CO—Cl + HCl}$
		$\mathrm{H_2N—R—CO_2H \rightarrow H{+}NH—R—CO{+}_nOH + H_2O}$
Polyester	—CO—O—	$\mathrm{HO—R—OH + HO_2C—R'—CO_2H \rightarrow H{+}O—R—OCO—R'—CO{+}OH + H_2O}$
		$\mathrm{HO—R—OH + CO_2R''—R'—CO_2R'' \rightarrow H{+}O—R—OCO—R'—CO{+}_nOH + R''OH}$
		$\mathrm{HO—R—CO_2H \rightarrow H{+}O—R—CO{+}_nOH + H_2O}$
Polyurethane	—O—CO—NH	$\mathrm{HO—R—OH + NCO—R'—NCO \rightarrow {+}O—R—OCO—NH—R'—NH—CO{+}_n + H_2O}$
Polysiloxane	—Si—O—	$\mathrm{Cl—SiR_2—Cl \xrightarrow[-HCl]{H_2O} HO—SiR_2—OH \rightarrow H{+}O—SiR_2{+}_nOH + H_2O}$
Phenol-formaldehyde	—Ar—$\mathrm{CH_2}$—	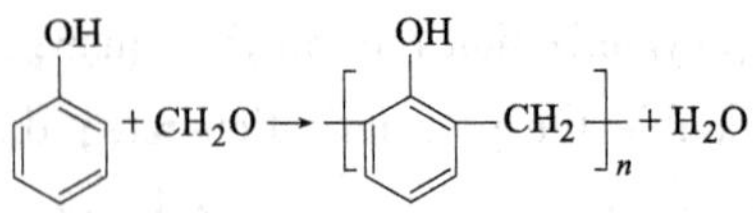

All step polymerizations fall into two groups depending on the type of monomers employed. The first involves two different bifunctional and/or polyfunctional monomers in which each monomer possesses only one type of functional group (A polyfunctional monomer is a monomer with more than one functional group per molecule. A bifunctional monomer is one with two functional groups oer molecule). The second involves a single monomer containing both types of functional groups. The synthesis of polyamides illustrates both groups of polymerization reactions. Thus polyamindes can be obtained from the reaction of diamides with diacids

$$nH_2N\text{—}R\text{—}NH_2 + nHO_2C\text{—}R'\text{—}CO_2H \longrightarrow$$
$$H\text{—}(NH\text{—}R\text{—}NHCO\text{—}R'\text{—}CO)_n\text{—}OH + (2n-1)H_2O \quad (7.21)$$

or from the reaction of amino acids with themselves:

$$nH_2N\text{—}R\text{—}COOH \longrightarrow H\text{—}(NH\text{—}R\text{—}CO)_n\text{—}OH + (n-1)H_2O \quad (7.22)$$

The two groups of reactions can be represented in a general manner by the equations:

$$nA\text{—}A + nB\text{—}B \longrightarrow \text{—}(A\text{—}AB\text{—}B)_n\text{—} \quad (7.23)$$
$$nA\text{—}B \longrightarrow \text{—}(A\text{—}B)_n\text{—} \quad (7.24)$$

where A and B are the two different types of functional groups. The characteristics of these two polymerization reactions are very similar. The successfully synthesis of high polymers using any step polymerization reaction generally is more difficult than the corresponding small molecule reaction, since high polymers can be achieved only at very high conversions (>98%). A conversion of 90%, which would be considered excellent for the synthesis of ethyl acetate or methyl benzamide, is a disaster for the synthesis of the corresponding polyester or polyamide. The need for very high conversions to synthesize high polymer places several stringent requirements on any reaction to be used for polymerization—a favorable equilibrium and the absence of cyclization and other side reactions. These stringent requirements are met by a relatively small fraction of the reactions familiar to and used by chemists to synthesize small molecules. An additional requirement for achieving high molecular weights is the need for A and B functional groups to be present at very close to stoichiometric amounts.

Consider the polyesterification of a diacid and a diol to illustrate the general form of the kinetics of a typical step polymerization. Simple esterification is a well-know acid catalyzed reaction and polyesterification follows the same course. The reaction involves proponation of the carboxylic acid,

$$\sim\!\!\sim C(=O)\text{—}OH + HA \underset{k_2}{\overset{k_1}{\rightleftharpoons}} \sim\!\!\sim \overset{+}{C}(OH)\text{—}OH(A^-) \quad (7.25)$$

I

followed by reaction of the protonated species I with the alcohol to yield the ester. In the above equations, ~~ is used to indicate all acid or alcohol species in the reaction mixture (i.e., monomer, dimer, trimer, ⋯ n-mer). Species I and II are shown in the form of their associated ion pairs since polymerization often takes place in organic media of low polarity. A^- is the negative counterion derived from the acid HA. Polyesterifications, like many other step polymerizations, are equilibrium reactions. However, from the practical viewpoint of obtaining high yields of high-molecule-weight product, such polymerizations are run in a manner so as to continuously shift the equilibrium in the direction of the polymer. In the case of a polyesterification, this is easily accomplished by removal of the water that is a by-product of the reaction species II [Eq. (7.26)]. Under these conditions the kinetics of polymerization can be handled by considering the reactions in Eqs. (7.26) and (7.27) to be irreversible.

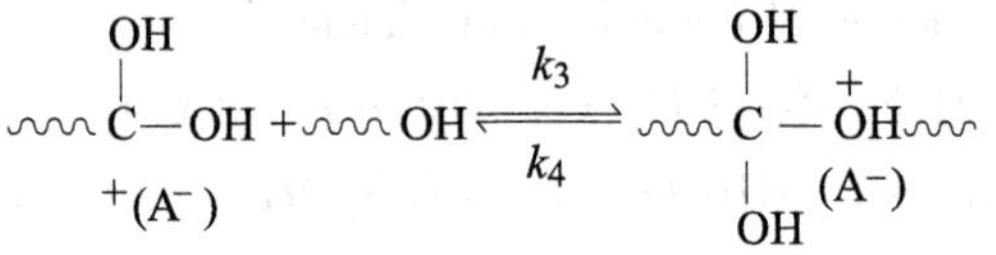

(7.26)

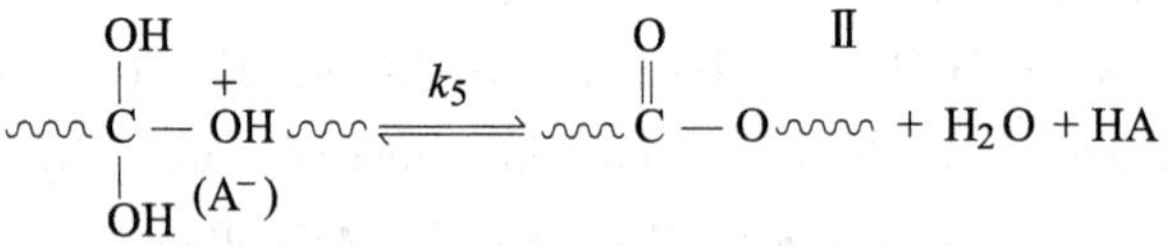

(7.27)

The rate of a step polymerization is conveniently expressed in terms of the concentrations of the reacting functional groups. Thus the polyesterification can be experimentally followed by titrating for the unreacted carboxyl groups with a base. The rate of polymerization R_p can then be expressed as the rate of disappearance of carboxyl groups $-d[COOH]/dt$. For the usual polyesterification, the polymerization rate is synonomous with the rate of formation of species II; that is, k_4 is vanishingly small[since the reaction is run under conditions that drive Eqs. (7.26) and (7.27) to the right], and k_1, k_2, and k_5 are large compared to k_3. An expression for the reaction rate can be obtained following the general procedures for handling a reaction scheme with the characteristics described. The rate of polymerization is given by

$$R = -\frac{d[COOH]}{dt} = k_3[\overset{+}{C}(OH)_2][OH] \tag{7.28}$$

where $[COOH]$, $[OH]$, and $[\overset{+}{C}(OH)_2]$ represent the concentrations of carboxyl, hydroxyl, and protonated carboxyl (I) groups, respectively. The concentration terms are in units of moles of the particular functional group per liter of solution.

Eq. (7.28) is in convenient in that the concentration of protonated carboxylic groups is not easily determined experimentally. One can obtain a more convenient expression for R_p by substituting for $[\overset{+}{C}(OH)_2]$ from the equilibrium expression

$$K = \frac{k_1}{k_2} = \frac{[\overset{+}{C}(OH)_2]}{[COOH][HA]} \tag{7.29}$$

for the protonation reaction Eq. (7.26). Combination of Eqs. (7.27) and (7.29) yields

$$-\frac{d[COOH]}{dt} = k_3K[COOH][HA][OH] \tag{7.30}$$

Two quite different kinetic situations arise from Eq. (7.30) depending on the identity of HA, that is, on whether a strong acid such as sulfuric acid or *p*-toluenesulfonic acid is added as an external catalyst.

7.2.1 Self-catalyzed Polymerization

In the absence of an externally added strong acid the diacid monomer acts as its own catalyst for the esterification reaction. For this case [HA] is replaced by [COOH] and Eq. (7.30) can be written in the usual form

$$-\frac{d[COOH]}{dt} = k_3K[COOH]^2[OH] \tag{7.31}$$

where K and k_3 have been combined into the experimentally determined rate constant k. Eq. (7.31) shows the important characteristic of the self-catalyzed polymerization—the reaction is third-order overall with a second-order dependence on the carboxyl concentration. The second-order dependence on the carboxyl concentration is comprised of two first-order dependencies—one for the carboxyl as the reactant and one as the catalyst. For most polymerizations the concentrations of the two functional groups are very nearly stoichiometric, and Eq. (7.31) can be written as

$$-\frac{d[M]}{dt} = k[M]^3 \tag{7.32}$$

or

$$-\frac{d[M]}{[M]^3} = kdt \tag{7.33}$$

where [M] is the concentration of hydroxyl groups or carboxyl groups. Integration of Eq. (7.33) yields

$$2kt = \frac{1}{[M]^2} - \frac{1}{[M]_0^2} \tag{7.34}$$

where $[M]_0$ is the initial (at $t=0$) concentration of hydroxyl or carboxyl groups. It is convenient at this point to write Eq. (7.34) in terms of the extent or fraction of reaction p defined as the fraction of the hydroxyl or carboxyl functional groups that has reacted at time t. p is also referred to as the extent or fraction of conversion (The value of p is calculated from a determination of the amount of unreacted carboxyl groups). The concentration [M] at time t of either hydroxyl or carboxyl groups is then given by

$$[M] = [M]_0 - [M]_0p = [M]_0(1-p) \tag{7.35}$$

Combination of Eqs. (7.34) and (7.35) yields

$$\frac{1}{(1-p)^2} = 2[M]_0^2kt + 1 \tag{7.36}$$

7.2.2 Molecular Weight of Polymer

The molecular weight of a polymer is of prime concern from the practical viewpoint, for unless a polymer is of sufficient high molecular weight (approximately >5000~10000), it will not have the desirable strength characteristics. It is therefore important to consider the change in polymer molecular weight with reaction time. For the case of stoichiometric amounts of diol and diacid, the number of unreacted carboxyl groups N is equal to the total number of molecules present in the system at some time t. This is so because each molecule larger than a monomer will on the average has a hydroxyl at one end and a carboxyl at the other end, while each diacid monomer molecule contains two carboxyls and each diol monomer contains no carboxyls.

The residue from each diol and each diacid (separately, not together) in the polymer chain is termed a structural unit (or a monomer unit). The repeating unit of the chain consists of two structural units, one each of the diol and diacid. The total number of structural units in any particular system equals the total number of bifunctional monomers initially present. The number-average degree of polymerization X_n is defined as the average number of structural units per polymer chain (The symbols P and DP are also employed to signify the number-average degree of polymerization). X_n is simply given as the total number of monomer molecules initially present divided by the total number of molecules present at time t:

$$\overline{X}_n = \frac{N_0}{N} = \frac{[M]_0}{[M]} \tag{7.37}$$

Combination Eqs. (7.35) and (7.37), one obtains

$$\overline{X}_n = \frac{1}{1-p} \tag{7.38}$$

This equation relating the degree of polymerization to the extent of reaction was originally set forth by Carothers and is often referred as the Carothers equation.

The number-average molecular weight M_n, defined as the total weight of a polymer sample divided by the total number of moles in it, is given by

$$\overline{M}_n = M_0 \overline{X}_n + M_{eg} = \frac{M_0}{1-p} + M_{eg} \tag{7.39}$$

where M_0 is the mean of the molecular weights of the two structural units, and M_{eg} is the molecular weight of the end groups. For the polyesterification of adipic acid, $HO_2C(CH_2)_4CO_2H$, and ethylene glycol, $HOCH_2CH_2OH$, the repeating unit is $OCH_2CH_2OCO(CH_2)_4CO$ and one half of its weight or 86 is the value of M_0. The end groups are H— and OH— and M_{eg} is 18. For even a modest molecular weight polymer the contribution of M_{eg} to M_n is negligibly small, and Eq. (7.39) becomes

$$\overline{M}_n = M_0 \overline{X}_n = \frac{M_0}{1-p} \tag{7.40}$$

Combination of Eqs. (7.36) and (7.40) yields

$$\overline{X}_n^2 = 1 + 2[M]_0^2 kt \tag{7.41}$$

Since the reaction time and degree of polymerization appear as the first and second powers, respectively, the polymer molecular weight will increase very slowly with reaction time except in the final stages of the reaction. This means that very long reaction times are needed to obtain a high-molecular-weight polymer product. The right-hand ordinate of Fig. 7.1 shows the variation of X_n and t. The slow increase of the molecular weight of the polymer with time is clearly apparent. The rate of increase of X_n with time decreases as the reaction proceeds. The production of high polymers requires reaction times that are too long from the practical viewpoint.

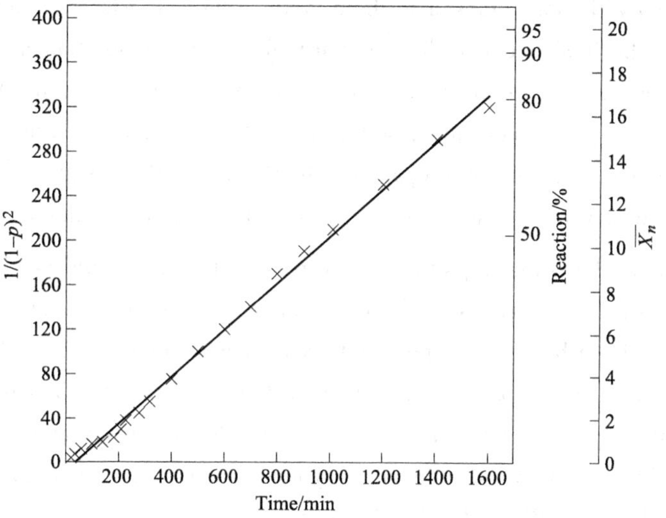

Fig. 7.1 Third-order plot of the self-catalyzed polyesterification of adipic acid with diethylene glycol at 166℃

7.3 Polymerization Processes

Bulk Polymerization Monomer and polymer (with traces of initiator) are the only constituents in bulk polymerizations. Obviously, the monomer must be soluble in the polymer for this type of process to effectively proceed. Bulk polymerization, also called mass or block polymerization, can occur in stirred-tank reactors, or can be unstirred, in which instance it is called quiescent bulk polymerization. The primary difficulty with bulk polymerization is that as the polymerization proceeds and more polymer is formed, the viscosity increases, thermal conductivity decreases, and heatremoval become difficult.

There is an increase in temperature as heat removal slows, along with a corresponding increase in the reaction rate. This phenomenon is known as the autoacceleration or Trommsdorff effect and can lead to catastrophic results if not properly controlled. Even with successful control of the reaction, it is difficult to remove the traces of remaining monomer from the polymer due to decreased diffusion. Similarly, it is difficult to get the reactions to proceed to completion due to limited monomer mobility.

A bulk polymerization reactor can be as simple as a tube into which the reactants are fed and from which the polymer mixture emerges at the end; it can be more of a traditional, continuous stirred-tank reactor (CSTR), or even a high-pressure autoclave-typereactor. A bulk polymerization process need not be continuous, but it should not be confused with a batch reaction. There can be batch bulk polymerizationsjust as there are continuous bulk polymerizations processes.

Solution Polymerization By adding a solvent to the monomer-polymermixture, heat removal can be improved dramatically over bulk reactions. The solvent must be removed after the polymerization is completed, however, which leads to a primary disadvantage of solution polymerization. Another problem associated with radical chain polymerizations carried out in solution is associated with chain transfer to the solvent. Chain transfer can significantly affect the molecular weight of the final polymer. This is particularly true in solution polymerization, where there are many solvent molecules present. In fact, chain transfer to solvent often dominates over chain transfer to other types of molecules, so that Eq. (7.42) reduces to where the first term once again describes the contributions of propagation to the molecular weight, [S] and [M] are concentrations of the solvent and monomer, respectively, and C_S is the ratio of rate constants $k_{tr,s}/k_p$. This equation shows that a plot of $1/\overline{X}_n$ versus [S]/[M] should yield a straight line, the intercept for which gives the degree of polymerization in the absence of chain transfer to solvent. This relationship has been borne out experimentally for polystyrene and several solvent systems.

The reactor equipment used for solution polymerizations is typically glass-lined stainless steel. An example of solution polymerization is the reaction of ethylene inisooctane with a chromia silica alumina catalyst initiator to form polyethylene. Typical reaction conditions for this polymerization are 150~180℃ and 2.1~4.8MPa (300~700 psi).

$$\frac{1}{\overline{X}_n} = \frac{\gamma f k_d k_t^{\frac{1}{2}}[I]}{k_p[M]} + \sum_i \left(\frac{k_{tr,i}[A]}{k_p[M]}\right) \tag{7.42}$$

$$\frac{1}{\overline{X}_n} = \left(\frac{1}{\overline{X}_n}\right)_0 + \frac{C_S[S]}{[M]} \tag{7.43}$$

Suspension Polymerization When water is the solvent and the monomer is insoluble in water, suspension polymerization is often employed. The monomer and initiator form small droplets within the aqueous phase, and polymerization proceeds via bulk polymerization. Droplet sizes are typically between 10 and 1000μm in diameter, so that heat transfer to the suspending medium and subsequently to the reactor walls is very efficient throughout the course of the reaction. The result of the polymerization within each of the droplets is a polymer "bead". Though some agglomeration of particles takes place, the addition of a suspending agent such as poly (vinylalcohol) minimizes coalescence, and the final polymer bead diameters are typically 100~1000μm. The polymer beads are recovered by filtration, which is followed by a washing step. Suspension polymerization can be carried to nearly 90% completion at moderate pressures of 600kPa (88 psi) and 80℃ with stirring at about 150 to 300 rpm.

Emulsion Polymerization Emulsion polymerization is similar to suspension polymerization in that the monomer is insoluble in the solvent, water. The similarity ends there. In emulsion polymerization, the initiator is water-soluble, the particles in which the propagation step occurs are smaller, typically 0.05~5μm in diameter, and the final beads are not usually filtered from the aqueous solution. The emulsifier, or surfactant, plays a very important role in emulsion polymerization. It is usually a long-chained hydrocarbon with a hydrophylic "head" and hydrophobic "tail". This charge polarization causes the tails to surround the monomer molecules, with the heads pointing outward into the water phase, forming a spherical droplet called a micelle (see Fig. 7.2). As radicals initiate in the aqueous phase, they react with what few monomer molecules are dissolved in the water. As the radicals slowly grow and become more hydrophobic, they find it more stable to be inside of the micelles. Upon entering a micelle, the radicals cause rapid propagation and the micelle expands.

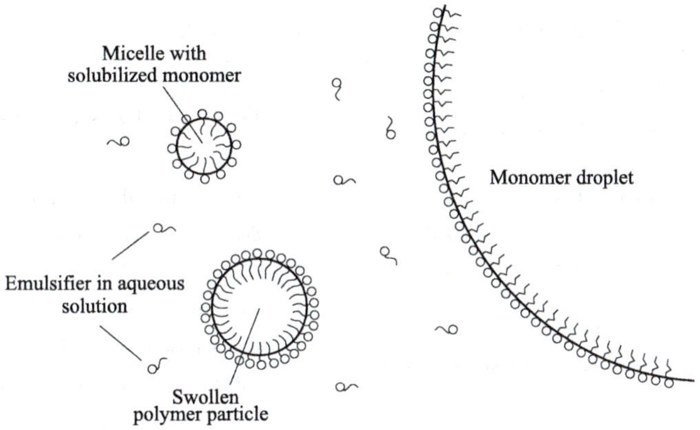

Fig. 7.2 Schematic representation of micelle formation in emulsion polymerization

It can be shown that the degree of polymerization in emulsion polymerization can be derived from general polymerization kinetics [Eq. (7.42)] by neglecting chain transfer:

$$\frac{1}{\overline{X}_n}=\frac{2fk_d[\mathrm{I}]}{k_p[\mathrm{M}^*][\mathrm{N}^*]} \tag{7.44}$$

where the superscript " * " on the monomer concentration indicates that this is the concentration of monomer in the micelle, not the solution. Since the initiator is solublen water, its concentration remains based upon the total volume of water present. In Eq. (7.44), $[\mathrm{N}^*]$ is the concentration of micelle-swollen polymer particles.

The final product of emulsion polymerization is an emulsion—astable, heterogeneous mixture of fine polymer beads in an aqueous solution, sometime called a latex emulsion. Water-based paints, for example, can be formed from the emulsion polymerization of vinyl acetate. In this process, 1m^3 of water containing 3% poly (vinylalcohol) and 1% surfactant are heated to 60℃ in a reaction vessel. The temperature rises to around 80℃ over a 4 to 5 hour period as monomer and anaqueous persulfate solution are added. The rate at which heat can be removed limits the rate at which monomer can be added.

Copolymerizations can also be carried out in an emulsion reaction. For example, nitrile rubber (NBR) can be made from the emulsion polymerization of acrylonitrile and butadiene:

$$\underset{\text{Acrylonitrile}}{H_2C{=}C{-}CN} + \underset{\text{Butadiene}}{H_2C{=}CH{-}HC{=}CH_2} \longrightarrow \left[H_2C{-}\underset{\substack{|\\CN}}{CH}{-}CH_2{-}CH{=}CH{-}CH_2 \right]_n \tag{7.45}$$

As the acrylonitrile content in nitrile rubber increases, so does the resistance to nonpolar solvents.

The comparison of polymerization processes are shown in Table 7.2.

Table 7.2 Comparison of polymerization processes

Type	Advantages	Disadvantages
Batch bulk	Minimum contamination Simple equipment	Poor heat control Broad M_w distribution
Continuous bulk	Better heat control Narrow M_w distribution	Requires stirring, separation, and recycle
Solution	Good heat control Direct use of solution possible	Not useful for dry polymer Solvent removal difficult
Suspension	Excellent heat control Direct use of suspension possible	Requires stirring Contamination by stabilizer possible Additional processing (washing, drying) required
Emulsion	Excellent heat control Narrow M_w distribution High M_w attainable Direct use of emulsion possible	Contamination by emulsifier likely Additional processing (washing, drying) required

Questions

1. Describe and draw the structure of the polyester obtained in each of the following polymerizations:

   a. HO—R—COOH

   b. HO—R—COOH+HO—R′—OH

   c. HO—R—COOH+HO—R″—OH
   (with OH attached to R″)

   d. HO—R—COOH+HO—R′—OH+HO—R″—OH
   (with OH attached to R″)

   Would the structure of the polymer produced in each of cases b, c, and d depend on the relative amounts of the reactants? If so, describe the differences.

2. Show by chemical equations the polymerization of acrylonitrile initiated by the thermal decomposition of cumyl hydroperoxide.

3. For a radical polymerization with bimolecular termination, the polymer produced contains 1.3 initiator fragments per polymer molecule. Calculate the relative extents of termination by disproportionation and coupling, assuming that no chain-transfer reactions occur.

4. What are the characteristic overall features that distinguish emulsion polymerization from homogenous polymerization?

5. Describe the microscopic picture of emulsion polymerization according to Harkins, Smith, and Ewart. Where are the monomer, initiator, and emulsifier located? Describe the changes that take place as the reaction proceeds to 100% conversion.

References

[1] Li J, Yu Q, Wan T, et al. Introduction to polymer materials (bilingual teaching) [M]. Beijing: Chemical Industry Press, 2014.

[2] Brian S. Mitchell. An introduction to materials engineering and science[M]. Tulane University: A John Wiley & Sons, 2004: 255~261.

[3] Billmeyer F W. Textbook of polymer science[M]. Hoboken: Wiley, 1976.

Chapter 8 Corrosion and Protection of Materials

To one degree or another, most materials experience some type of interaction with a large number of diverse environments. Often, such interactions impair a material's usefulness as a result of the deterioration of its mechanical properties (e. g., ductility and strength), other physical properties, or appearance. Occasionally, to the chagrin of a design engineer, the degradation behavior of a material for some application is ignored; with adverse consequences. Deteriorative mechanisms are different for the three material types. In metals, there is actual material loss either by dissolution (corrosion) or by the formation of nonmetallic scale or film (*oxidation*). Ceramic materials are relatively resistant to deterioration, which usually occurs at elevated temperatures or in rather extreme environments; the process is frequently also called corrosion. For polymers, mechanisms and consequences differ from those for metals and ceramics, and the term degradation is most frequently used. Polymers may dissolve when exposed to a liquid solvent, or they may absorb the solvent and swell; also, electromagnetic radiation (primarily ultraviolet) and heat may cause alterations in their molecular structures. The deterioration of each of these material types is discussed in this chapter with special regard to mechanism, resistance to attack by various environments, and measure to prevent or reduce degradation.

Corrosion is defined as the destructive and unintentional attack on a metal; it is electrochemical and ordinarily begins at the surface. The problem of metallic corrosion is significant; in economic terms, it has been estimated that approximately 5% of an industrialized nation's income is spent on corrosion prevention and the maintenance replacement of products lost or contaminated as a result of corrosion reactions. The consequences of corrosion are all too common. Familiars' examples include the rusting of automotive body panels and radiator and exhaust components. Corrosion processes are occasionally used to advantage. For example, etching procedures use the selective chemical reactivity of grain boundaries or various micro-structural constituents.

8.1 Electrochemical Considerations

8.1.1 Electrode Potentials

When a current passes through the external circuit, electrons generated from the oxidation of iron flow to the copper cell in order that Cu^{2+} be reduced. In addition, there will be some net ion motion from each cell to the other across the membrane. This is called a galvanic couple—two metals electrically connected in a liquid electrolyte in which one metal becomes an anode and corrodes while the other acts as a cathode. An electric potential or voltage exists between the two cell halves, and its magnitude can be determined if a voltmeter is connected in the external circuit. A potential of 0.780V results for a copper-iron galvanic cell when the temperature is 25℃ (77℉).

Now consider another galvanic couple consisting of the same iron half-cell connected to a metal zinc electrode that is immersed in a 1M solution of Zn^{2+} ions. In this case, the zinc is the anode and corrodes, whereas the Fe becomes the cathode. The electrochemical reaction is thus

$$Fe^{2+} + Zn \longrightarrow Fe + Zn^{2+} \tag{8.1}$$

The potential associated with this cell reaction is 0.323V. Thus, various electrode pairs have different voltages; the magnitude of such a voltage may be thought of as representing the driving force for the electrochemical oxidation-reduction reaction. Consequently, metallic materials may be rated as to their tendency to experience oxidation when coupled to other metals in solutions of their respective ions. A half-cell similar to those just described [i.e., a pure metal electrode immersed in a 1M solution of its ions and at 25℃ (77℉)] is termed a standard half-cell.

8.1.2 The Standard EMF Series

These measured cell voltages represent only differences in electrical potential, and thus it is convenient to establish a reference point, or reference cell, to which other cell halves may be compared. This reference cell, arbitrarily chosen, is the standard hydrogen electrode. It consists of an inert platinum electrode in a 1M solution of H ions saturated with hydrogen gas that is bubbled through the solution at a pressure of 1 atm and a temperature of 25℃ (77℉). The platinum itself does not take part in the electrochemical reaction; it acts only as a surface on which hydrogen atoms may be oxidized or hydrogen ions may be reduced. The electromotive force (EMF) series is generated by coupling to the standard hydrogen electrode, standard half-cells for various metals and ranking them according to measured voltage.

8.1.3 The Galvanic Series

Even though Table 8.1 was generated under highly idealized conditions and has limited utility, it nevertheless indicates the relative reactivities of the metals. A more realistic and practical ranking is provided by the galvanic series, Table 8.1. This represents the relative reactivities of a number of metals and commercial alloys in seawater. The alloys near the top are cathodic and unreactive, whereas those at the bottom are most anodic; no voltages are provided. Comparison of the standard EMF and the galvanic series reveals a high degree of correspondence between the relative positions of the pure base metals. Most metals and alloys are subject to oxidation or corrosion to one degree or another in a wide variety of environments—that is, they are more stable in an ionic state than as metals. In thermodynamic terms, there is a net decrease in free energy in going from metallic to oxidized states. Consequently, essentially all metals occur in nature as compounds, for example, oxides, hydroxides, carbonates, silicates, sulfides, and sulfates. Two exceptions are the noble metals gold and platinum, for which oxidation in most environments is not favorable, and, therefore, they may exist in nature in the metallic state.

Table 8.1 The summary of units for the two corrosion penetration rate schemes

CPR Units	K Value	Units			
		W	ρ	A	t
mpy	534	mg	g/cm^3	in^2	h
mm/yr	87.6	mg	g/cm^3	cm^3	h

8.1.4 Corrosion Rates

The half-cell potentials listed in Table 8.1 are thermodynamic parameters that relate to systems at equilibrium. For example, for the discussions pertaining, it was tacitly assumed that there was no current flow through the external circuit. Real corroding systems are not at equilibrium; there is a flow of electrons from anode to cathode, which means that the half-cell potential parameters (Table 8.1) cannot be applied.

Furthermore, these half-cell potentials represent the magnitude of a driving force, or the tendency for the occurrence of the particular half-cell reaction. However, although these potentials may be used to determine spontaneous reaction directions, they provide no information on corrosion rates. That is, even though a V potential computed for a specific corrosion situation using Eq. (8.1) is a relatively large positive number, the reaction may occur at only an insignificantly slow rate. From an engineering perspective, we are interested in predicting the rates at which systems corrode; this requires the use of other parameters, as discussed next. The corrosion rate, or the rate of material removal as a consequence of the chemical action, is an important corrosion parameter. This may be expressed as the corrosion penetration rate (CPR), or the thickness loss of material per unit of time. The formula for this calculation is

$$\text{CPR} = KW/\rho At \tag{8.2}$$

where W is the weight loss after exposure time t; ρ and A represent the density and exposed specimen area, respectively, and K is a constant, its magnitude depending on the system of units used.

8.1.4.1 Polarization

Consider the standard Zn/H_2 electrochemical cell, which has been short-circuited such that oxidation of zinc and reduction of hydrogen occurs at their respective electrode surfaces. The potentials of the two electrodes are not at the values determined because the system is now a nonequilibrium one. The displacement of each electrode potential from its equilibrium value is termed polarization, and the magnitude of this displacement is the *overvoltage*, normally represented by the symbol η. Overvoltage is expressed in terms of plus or minus volts (or millivolts) relative to the equilibrium potential. For example, suppose that the zinc electrode in Fig. 8.1 has a potential of 0.621V after it has been connected to the platinum electrode. The equilibrium potential is 0.763V, therefore,

$$\eta = -0.621\text{V} - (-0.763\text{V}) = +0.142\text{V} \tag{8.3}$$

There are two types of polarization—activation and concentration. We now discuss their mechanisms because they control the rate of electrochemical reactions.

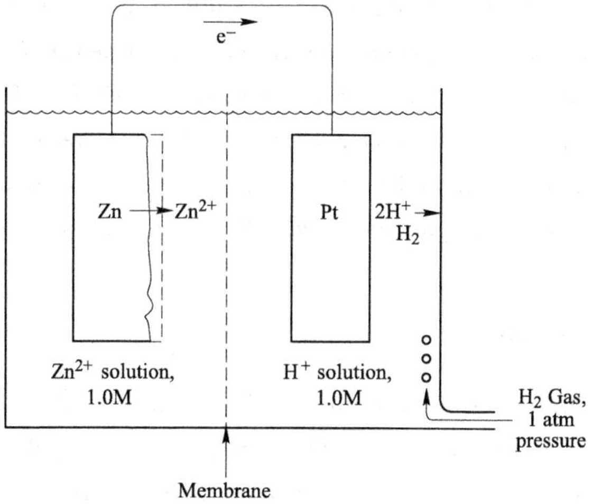

Fig. 8. 1 Electrochemical cell consisting of standard zinc and hydrogen electrodes that has been short-circuited

8. 1. 4. 2 Passivity

Under particular environmental conditions, some normally active metals and alloys, lose their chemical reactivity and become extremely inert. This phenomenon, termed passivity, is displayed by chromium, iron, nickel, titanium, and many of their alloys. It is believed that this passive behavior results from the formation of a highly adherent and very thin oxide film on the metal surface, which serves as a protective barrier to further corrosion. Stainless steels are highly resistant to corrosion in a rather wide variety of atmospheres as a result of passivation. They contain at least 11% chromium, which as a solid solution alloying element in iron, minimizes the formation of rust; instead, a protective surface film forms in oxidizing atmospheres (Stainless steels are susceptible to corrosion in some environments and therefore are not always "stainless") . Aluminum is highly corrosion resistant in many environments because it also passivates. If damaged, the protective film normally re-forms very rapidly. However, a change in the character of the environment (e. g., alteration in the concentration of the active corrosive species) may cause a passivated material to revert to an active state. Subsequent damage to a preexisting passive film could result in a substantial increase in corrosion rate, by as much as 100000 times. This passivation phenomenon may be explained in terms of polarization potential-log current density curves discussed in the preceding section. The polarization curve for a metal that passivates has the general shape. At relatively low potential values, within the "active" region the behavior is linear, as it is for normal metals. Within creasing potential, the current density suddenly decreases to a very low value that remains independent of potential; this is termed the "passive" region. Finally, at even higher potential values, the current density again increases with potential in the "transpassive" region. The figure illustrates how a metal can experience both active and passive behavior depending on the corrosion environment. This figure shows the S-shaped oxidation polarization curve for an active-passive metal M and, in addition, reduction polarization curves for two different solutions, which are labeled 1 and 2. Curve 1 intersects the oxidation polarization curve in the active region at point A, yielding a corrosion current density $iC(A)$. The intersection of curve 2 at point B is in the passive region and at current density $iC(B)$. The corrosion rate of metal M in solution 1 is greater than in solution 2 because $iC(A)$ is greater than $iC(B)$ and rate is proportional to current density. This difference in corrosion rate between the two solutions may be significant—several orders of magnitude—given that the current density scale is scaled logarithmically.

8.2 Forms of Corrosion

It is convenient to classify corrosion according to the manner in which it is manifest. Metallic corrosion is sometimes classified into eight forms: uniform, galvanic, crevice, pitting, intergranular, selective leaching, erosion-corrosion, and stress corrosion. The causes and means of prevention of each of these forms are discussed briefly. In addition, we have elected to discuss the topic of hydrogen embrittlement in this section. Hydrogen embrittlement is, in a strict sense, a type of failure rather than a form of corrosion; however, it is often produced by hydrogen that is generated from corrosion reactions.

8.2.1 Uniform Attack

Uniform attack is a form of electrochemical corrosion that occurs with equivalent intensity over the entire exposed surface and often leaves behind a scale or deposit. Ina microscopic sense, the oxidation and reduction reactions occur randomly over the surface. Familiar examples include general rusting of steel and iron and the tarnishing of silverware. This is probably the most common form of corrosion. It is also the least objectionable because it can be predicted and designed for with relative ease.

8.2.2 Galvanic Corrosion

Galvanic corrosion occurs when two metals or alloys having different compositions are electrically coupled while exposed to an electrolyte. This is the type of corrosion or dissolution that was described in Section 8.2. The less noble or more reactive metal in the particular environment experiences corrosion; the more inert metal, the cathode, is protected from corrosion. As examples, steel screws corrode when in contact with brass in a marine environment, and if copper and steel tubing are joined in a domestic water heater, the steel corrodes in the vicinity of the junction. Depending on the nature of the solution, one or more of the reduction reactions, occurs at the surface of the cathode material.

8.2.3 Crevice Corrosion

Electrochemical corrosion may also occur as a consequence of concentration differences of ions or dissolved gases in the electrolyte solution and between two regions of the same metal piece. For such a *concentration cell*, corrosion occurs in the locale that has the lower concentration. A good example of this type of corrosion occurs in crevices and recesses or under deposits of dirt or corrosion products where the solution becomes stagnant and there is localized depletion of dissolved oxygen. Corrosion preferentially occurring at these positions is called crevice corrosion as shown in Fig. 8.2. The crevice must be wide enough for the solution to penetrate yet narrow enough for stagnancy; usually the width is several thousandths of an inch.

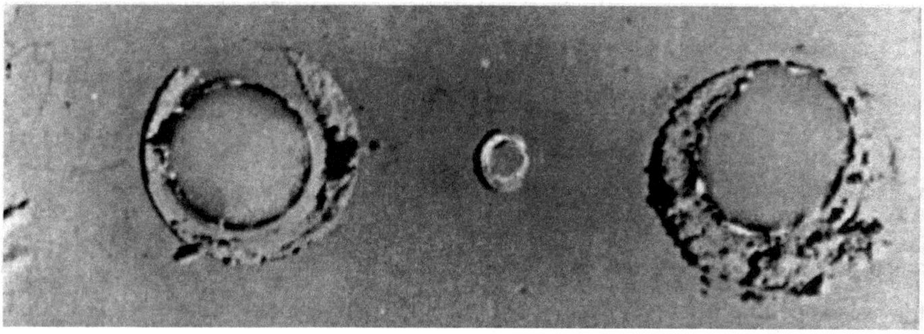

Scan the QR code to see color picture

Fig. 8.2 On this plate, which was immersed in seawater, crevice corrosion has occurred at the regions that were covered by washers
(Photograph courtesy of LaQue Center for Corrosion Technology, Inc)

The proposed mechanism for crevice corrosion is illustrated in Fig. 8.3. After oxygen has been depleted within the crevice, oxidation of the metal occurs at this position according to Eq. (8.5). Electrons from this electrochemical reaction are conducted through the metal to adjacent external regions, where they are consumed by reduction—most probably reaction (8.6). In many aqueous environments, the solution within the crevice has been found to develop high concentrations of H and Cl ions, which are especially corrosive. Many alloys that passivate are susceptible to crevice corrosion because protective films are often destroyed by the H and Cl ions. Crevice corrosion may be prevented by using welded instead of riveted or bolted joints, using non absorbing gaskets when possible, removing accumulated deposits frequently, and designing containment vessels to avoid stagnant areas and ensure complete drainage.

8.2.4 Pitting

Pitting is another form of much localized corrosion attack in which small pits or holes form. They ordinarily penetrate from the top of a horizontal surface downward in a nearly vertical direction. It is an extremely insidious type of corrosion, often going undetected and with very little material loss until failure occurs.

The mechanism for pitting is probably the same as for crevice corrosion, in that oxidation occurs within the pit itself, with complementary reduction at the surface. It is supposed that gravity causes the pits to grow downward, the solution at the pit tip becoming more concentrated and dense as pit growth progresses. A pit may be initiated by a localized surface defect such as a scratch or a slight

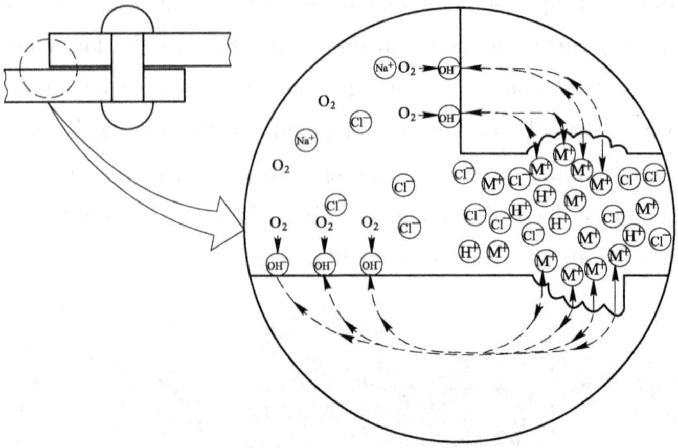

Fig. 8. 3 Schematic illustration of the mechanism of crevice corrosion between two riveted sheets

variation in composition. In fact, it has been observed that specimens having polished surfaces display a greater resistance to pitting corrosion. Stainless steels are somewhat susceptible to this form of corrosion; however, alloying with about 2% molybdenum enhances their resistance significantly.

8. 2. 5 Intergranular Corrosion

As the name suggests, intergranular corrosion occurs preferentially along grain boundaries for some alloys and in specific environments. The net result is that a macroscopic specimen disintegrates along its grain boundaries. This type of corrosion is especially prevalent in some stainless steels. When heated to temperatures between 500℃ and 800℃ (950℉ and 1450℉) for sufficiently long time periods, these alloys become sensitized to intergranular attack. It is believed that this heat treatment permits the formation of small precipitate particles of chromium carbide ($Cr_{23}C_6$) by reaction between the chromium and carbon in the stainless steel. These particles form along the grain boundaries. Both the chromium and the carbon must diffuse to the grain boundaries to form the precipitates, which leaves a chromium-depleted zone adjacent to the grain boundary. Consequently, this grain boundary region is now highly susceptible to corrosion.

Intergranular corrosion is an especially severe problem in the welding of stainless steels, when it is often termed weld decay. Fig. 8. 4 shows this type of intergranular corrosion. Stainless steels may be protected from intergranular corrosion by the following measures: (1) subjecting the sensitized material to a high-temperature heat treatment in which all the chromium carbide particles are re-dissolved; (2) lowering the carbon content below 0. 03%, so that carbide formation is minimal; (3) alloying the stainless steel with another metal such as niobium or titanium, which has a greater tendency to form carbides than does chromium so that the Cr remains in solid solution. Because only a porous mass of copper remains in the region that has been dezincified. Inaddition, the material changes from yellow to a red or copper color. Selective leachingmay also occur with other alloy systems in which aluminum, iron, cobalt, chromium, andother elements are vulnerable to preferential removal.

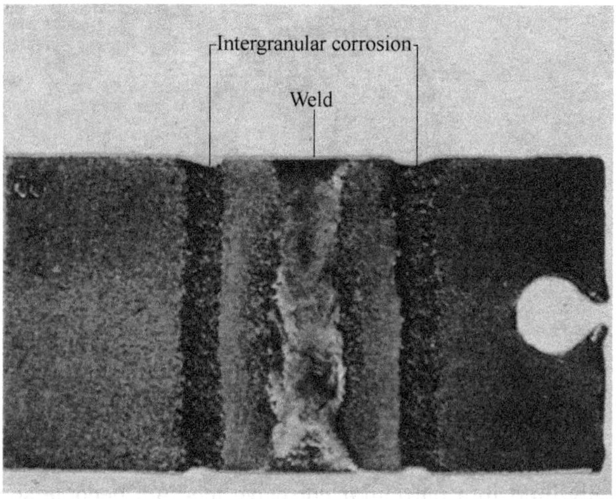

Fig. 8.4 Weld decay in a stainless steel
(The regions along which the grooves have formed were sensitized as the weld cooled)

8.2.6 Erosion-corrosion

Erosion-corrosion arises from the combined action of chemical attack and mechanical abrasion or wear as a consequence of fluid motion. Virtually all metal alloys, to one degree or another, are susceptible to erosion-corrosion. It is especially harmful to alloys that passivate by forming a protective surface film; the abrasive action may erode away the film, leaving exposed a bare metal surface. If the coating is not capable of continuously and rapidly re-forming as a protective barrier, corrosion may be severe. Relatively soft metals such as copper and lead are also sensitive to this form of attack. Usually it can be identified by surface grooves and waves having contours that are characteristic of the flow of the fluid. The nature of the fluid can have a dramatic influence on the corrosion behavior.

Increasing fluid velocity normally enhances the rate of corrosion. Also, a solution is more erosive when bubbles and suspended particulate solids are present. Erosion-corrosion is commonly found in piping, especially at bends, elbows, and abrupt changes in pipe diameter-positions where the fluid changes direction or flow suddenly becomes turbulent. Propellers, turbine blades, valves, and pumps are also susceptible to this form of corrosion. Fig. 8.5 illustrates the impingement failure of an elbow fitting. One of the best ways to reduce erosion-corrosion is to change the design to eliminate fluid turbulence and impingement effects. Other materials may also be used that inherently resist erosion. Furthermore, removal of particulates and bubbles from the solution lessens its ability to erode.

8.2.7 Stress Corrosion

Stress corrosion, sometimes termed stress corrosion cracking, results from the combined action of an applied tensile stress and a corrosive environment; both influences are necessary. In fact, some

Fig. 8. 5 Impingement failure of an elbow that was part of a steam condensate line

materials that are virtually inert in a particular corrosive medium become susceptible to this form of corrosion when a stress is applied. Small cracks form and then propagate in a direction perpendicular to the stress with the result that failure may eventually occur as depicted in Fig. 8. 6. Failure behavior is characteristic of that for a brittle material, even though the metal alloy is intrinsically ductile. Furthermore, cracks may form at relatively low stress levels, significantly below the tensile strength. Most alloys are susceptible to stress corrosion in specific environments, especially at moderate stress levels. For example, most stainless steels stress corrodes in solutions containing chloride ions, whereas brasses are especially vulnerable when exposed to ammonia. Fig. 8. 7 is a photomicrograph showing an example of intergranular stress corrosion cracking in brass. The stress that produces stress corrosion cracking need not be externally applied; it may be a residual one that results from rapid temperature changes and uneven contraction or occur for two-phase alloys in which each phase has a different coefficient of expansion. Also, gaseous and solid corrosion products that are entrapped internally can give rise to internal stresses. Probably the best

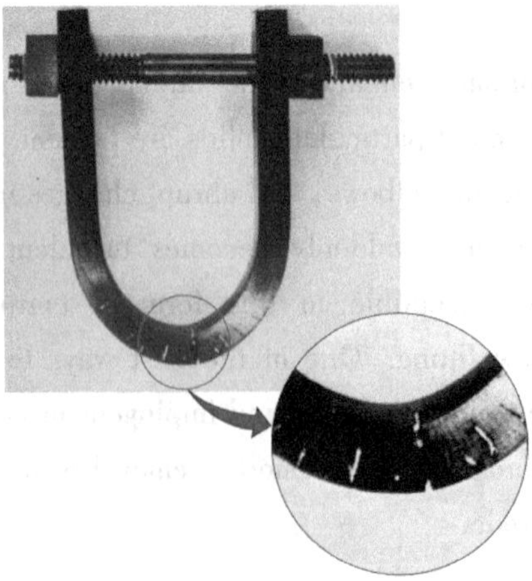

Fig. 8. 6 A bar of steel bent into a horseshoe shape using a nutand-bolt assembly

(While immersed in seawater, stress corrosion cracks formed along the bend at those regions where the tensile stresses are the greatest)

measure to take to reduce or completely eliminate stress corrosion is to lower the magnitude of the stress. This may be accomplished by reducing the external load or increasing the cross-sectional area perpendicular to the applied stress. Furthermore, an appropriate heat treatment may be used to anneal out any residual thermal stresses.

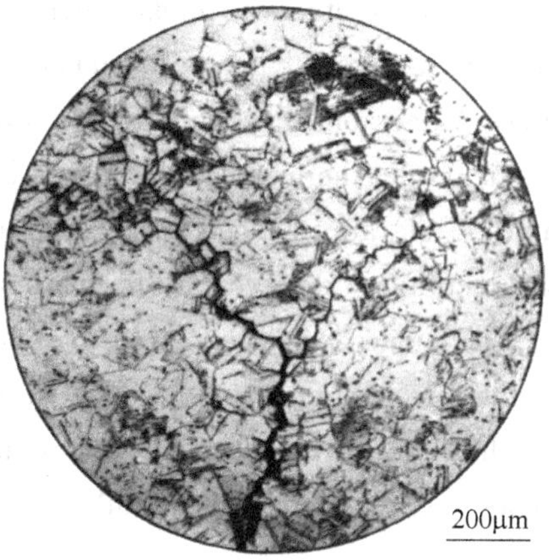

Fig. 8.7 Photomicrograph showing intergranular stress corrosion cracking in brass

8.2.8 Hydrogen Embrittlement

Various metal alloys, specifically some steels, experience a significant reduction in ductility and tensile strength when atomic hydrogen (H) penetrates into the material. This phenomenon is aptly referred to as hydrogen embrittlement; the terms *hydrogeninduced cracking and hydrogen stress cracking* are sometimes also used. Strictly speaking, hydrogen embrittlement is a type of failure; in response to applied or residual tensile stresses, brittle fracture occurs catastrophically as cracks grow and rapidly propagate.

Hydrogen in its atomic form (H as opposed to the molecular form, H_2) diffuses interstitially through the crystal lattice and concentrations as low as several parts per million can lead to cracking. Furthermore, hydrogen-induced cracks are most often transgranular, although intergranular fracture is observed for some alloy systems. A number of mechanisms have been proposed to explain hydrogen embrittlement; most are based on the interference of dislocation motion by the dissolved hydrogen. Hydrogen embrittlement is similar to stress corrosion, in that a normally ductile metal experiences brittle fracture when exposed to both a tensile stress and a corrosive atmosphere. However, these two phenomena may be distinguished on the basis of their interactions with applied electric currents. Whereas cathodic protection reduces or causes a cessation of stress corrosion, it may, however, lead to the initiation or enhancement of hydrogen embrittlement. For hydrogen embrittlement to occur, some source of hydrogen must be present, as well as the possibility for the formation of its atomic species. Situations in which these conditions are met include the following: pickling of steels in sulfuric acid; electroplating; and the presence of hydrogen-bearing atmospheres (including water vapor) at elevated temperatures such as during welding and heat treatments. Also, the presence of what are termed *poisons* such as sulfur (i. e., H_2S) and arsenic compounds accelerates hydrogen embrittlement; these substances retard the formation of molecular hydrogen and thereby increase the residence time of atomic hydrogen on the metal surface. Hydrogen sulfide, probably the most aggressive poison, is found in petroleum fluids, natural gas, oil-well brines, and geothermal fluids. High-strength steels are susceptible to hydrogen embrittlement, and increasing strength tends to enhance the material's susceptibility. Martensitic steels are especially vulnerable to this type of failure; bainitic, ferritic, and spheroiditic steels are more resilient. Furthermore, FCC alloys (austenitic stainless steels and alloys of copper, aluminum, and nickel) are relatively resistant to hydrogen embrittlement, mainly because of their inherently high ductilities. However, strain hardening these alloys enhances their susceptibility to embrittlement. Techniques commonly used to reduce the likelihood of hydrogen embrittlement include reducing the tensile strength of the alloy via a heat treatment, removing the source of hydrogen, "baking" the alloy at an elevated temperature to drive out any dissolved hydrogen, and substituting a more embrittlement-resistant alloy.

8.3 Protection Principles and Ways

Four different main principles can be used to prevent corrosion:

(1) Electrochemical, i. e. cathodic and anodic protection.

(2) Application of coatings.

(3) Change of environment.

(4) Suitable design.

The choice between these possibilities is usually based upon economic considerations, but in many cases aspects such as appearance, environment and safety must also be taken care of. Two or more of the four principles are commonly used at the same time. It is important to decide upon corrosion prevention at the design stage.

8.3.1 Cathodic Protection

Cathodic protection **(CP)** has been known for about 170 years. Primarily it has been used for protection of ordinary structural steel in soil and seawater, more seldom (and under special conditions) for steel exposed to fresh water. Other materials can also be protected by CP, for instance to prevent localized corrosion on stainless steel and aluminium. CP has in most cases been applied in combination with a coating, with the intention to protect the steel on "holidays" and damaged areas of the coating. In recent decades the application of this technology has increased considerably in connection with the expanding offshore oil and gas exploration and production. It is the dominating protection method for the large submerged parts of fixed oil installations, most often the only one for the parts freely exposed to seawater.

8.3.1.1 Principle

The main principle of CP is to impress an external current on the material, which forces the electrode potential down to the immune region, or, for protection against localized corrosion, below a protection potential. In other words, the material is made the cathode in an electrochemical cell; it is polarized cathodically as illustrated by potential-log current curves in Fig. 8.8.

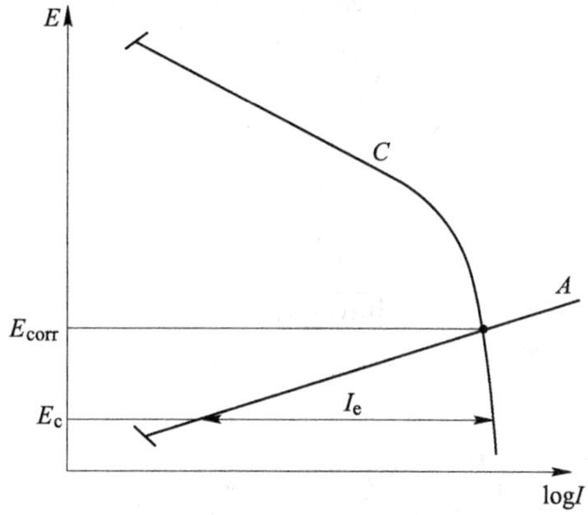

Fig. 8.8 Shift of potential from the free corrosion potential E_{corr} to a lower potential E_c by means of an external current I_e

The external current I_e is the difference between the cathodic and the anodic current at the actual potential E_c. At this potential, the anodic current on the protected material is very low, i. e. the external current is nearly equal to the cathodic current.

The external current can be produced in two different ways (Fig. 8.9):

(1) By means of a less noble material in the form of sacrificial anodes, which are connected by metallic conductors to the structure to be protected?

(2) By means of an external current source, usually a rectifier. A reference electrode may be used to control the rectifier potentiostatically.

A potential-log current diagram for a system with sacrificial anodes is shown in Fig. 8.10. The flat anodic curve of the sacrificial anode is typical for a suitable anode material. There is a potential difference ΔE between the structure and the anode due to a potential drop in the water. When the structure is completely protected, the galvanic current $I_g = I_c = I_a$, where I_c is the cathodic current on the structure, and I_a is the anodic current on the sacrificial anode, implying that the relatively small cathodic current on the anode is disregarded.

When dealing with CP we operate with different definitions of the term potential. We have the electrode potentials of the structure and the sacrificial anode, i. e. the cathode and the anode potential, and we have a varying electrical potential in the system. To avoid confusion as regards the sign, plus or minus, of potential differences from one place to another it is useful to present the electrical potential variation in a potential-distance diagram. For a CP system with sacrificial anodes, such a diagram is shown in Fig. 8.11. The system forms a galvanic element, and similar diagrams are also useful in connection with other cases of galvanic corrosion. In Fig. 8.11, any potential drop in metallic materials (in the structure and the connecting conductors) is disregarded since it is usually insignificant in comparison with the other potential differences. Therefore, the

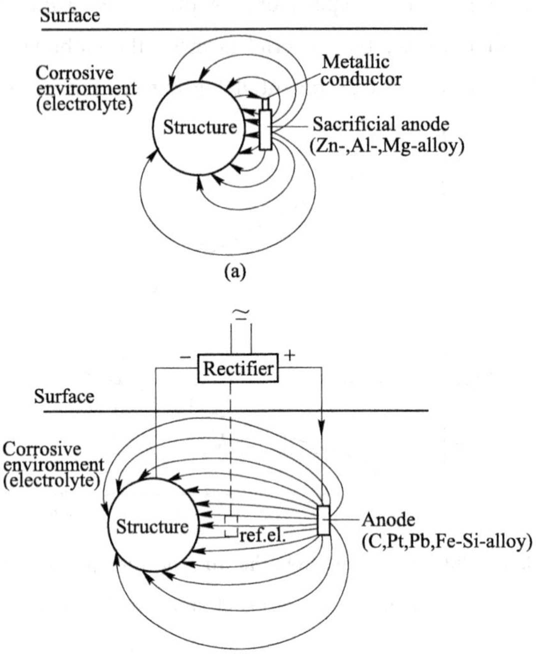

Fig. 8. 9 Cathodic protection

(a) Sacrificial anodes; (b) Impressed current

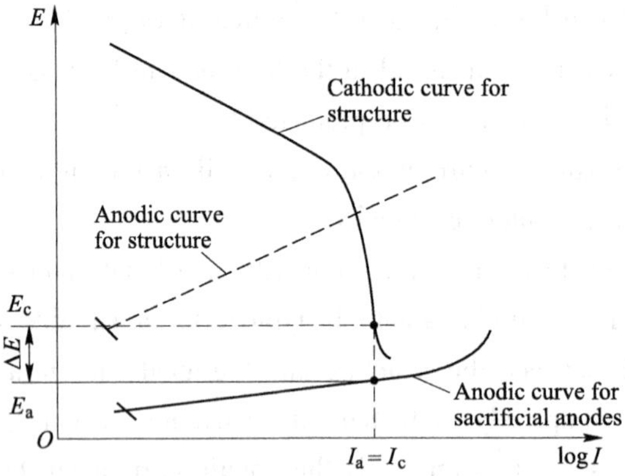

Fig. 8. 10 Potential-current diagram for cathodic protection with sacrificial anodes

presentation involves a common electrialc potential P_m in the anode and the structure material. The electrode potential is the difference in electrical potential between the metal and the adjacent electrolyte. Versus common reference electrodes such as the standard hydrogen electrode, the saturated calomel electrode and the silver-silver chloride electrode, the electrode potentials on the steel structure and the sacrificial anodes (E_c and E_a) are negative, as illustrated in the diagram. The anode potential E_a is more negative than the electrode potential of the structure E_c, due to a potential drop ΔE in the water. It should also be emphasized that the electrical potential in the

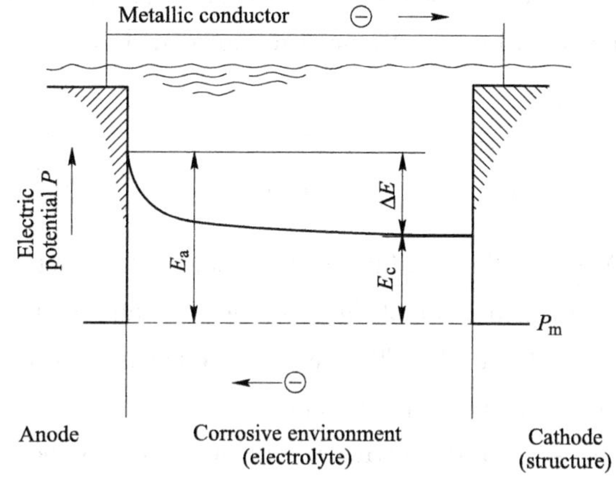

Fig. 8.11 Schematic potential-distance diagram for cathodic protection with sacrificial anodes

water decreases in the direction from the anode to the structure (positive ions move towards the structure, negative ions towards the anode). Most of the potential drop in the water is localized close to the anode because the surface of the anode in reality is much smaller than that of the structure, i. e. the electrolytic current density is much higher near the anode than elsewhere.

If an external current source is connected with its negative terminal to the structure an positive terminal to the anode, we obtain a protection system with impressed current. The potential distance diagram for this case is shown in Fig. 8.12.

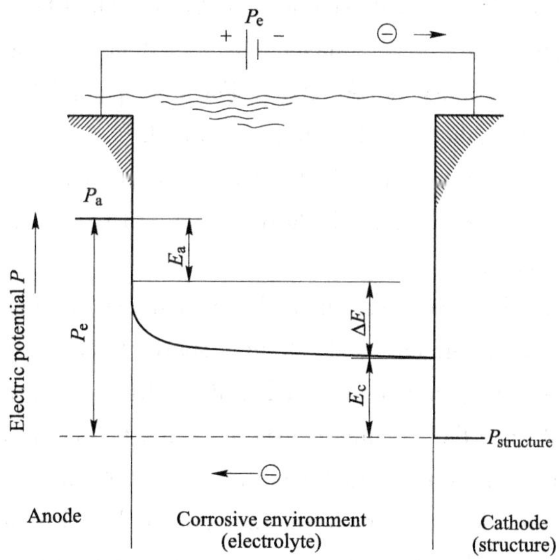

Fig. 8.12 Schematic potential-distance diagram for cathodic protection with impressed current

8.3.1.2 Protection Criteria and Specifications

In most cases it is appropriate to use a potential criterion for CP. The main criterion is $E_c < E_p$, where E_p is a protection potential defined mainly upon an empirical basis. For steel in seawater, the following protection potential has usually been recommended, referred to the various reference electrodes:

$Cu/CuSO_4$	Ag/AgCl/seawater	Zn	SCE	SHE
-850mV	-800mV	-250mV	-780mV	-530mV

This potential is 90mV above the equilibrium potential of the reaction $Fe^{2+} + 2e = Fe$ at an activity of $Fe^{2+} = 10^{-6}$. The stated value of the protection potential corresponds to the equilibrium potential at a ferrous ion activity of 10^{-3}. Theoretically, there is some corrosion at this potential, but experience shows that it is practically insignificant.

For steel buried in the seabed the corrosive environment may be affected by H_2S that is developed due to sulphate-reducing bacteria. Under these conditions the corrosion rate at the given value of E_p is higher, and therefore $E_p = -900$mV Reference Ag/AgCl is recommended in this case.

With a properly designed CP system with sacrificial anodes of a Zn- or Al-alloy, usually a potential in the range −900 to −1000mV is obtained after short time of exposure of a steel structure. This potential range also gives the lowest current consumption because a layer of calcareous deposits (dominated by $CaCO_3$) is soon formed. If an impressed current CP system is used, the potential may be considerably lower. Then, in addition to reduction of oxygen there will also be some hydrogen development. In certain cases this may cause damage.

(1) Cathodic disbonding of paint coatings due to hydroxide ions (OH^-) and possibly hydrogen gas produced in the cathodic reactions. Common paint systems for marine applications usually tolerate potentials down to −1100mV vs. Ag/AgCl. For lower potentials, a special documentation of the system's resistance to cathodic disbonding is required.

(2) Hydrogen embrittlement due to absorbed hydrogen atoms produced in the cathodic reaction. The risk for this type of damage exists particularly at high tensile stresses and for materials with high tensile strength and hardness.

Special conditions for welding, cold working and maximum material hardness ($HV_{10} = 350$) are specified. For materials with a minimum yield strength above 550MPa, qualification testing should be carried out when they are going to be used under critical conditions (e. g. at potentials <−1100mV Ag/AgCl and when the structure is exposed to plastic deformation and/or fatigue load) or when sufficient experience from the actual conditions does not exist.

To check that the current density is high enough, potential measurements have to be carried out, with reference electrodes placed (permanently or at regular monitoring) on potentially critical sites.

A general requirement is that conductors are thoroughly fastened to the anodes and the structure.

For CP of steel in seawater, Zn and Al anodes give sufficient driving voltage. For CP in environments with low conductivity (soil, fresh water), however, either impressed current or Mg sacrificial anodes are usually necessary.

8. 3. 1. 3 Cathodic Protection with Sacrificial Anodes

The majority of offshore installations, particularly in aggressive environments such as in the North Sea, are protected with sacrificial anodes. On submerged parts of fixed platforms, this is in most cases the only protection system. A few oil companies have used a combination of a paint coating system and sacrificial anodes, while a combination of thermally sprayed aluminium and sacrificial anodes have been used in special cases. Combinations of coatings and CP and/or a combination of sacrificial anodes and impressed current may be of particular interest for the reduction of weight and drag forces from the anodes on structures in deep water.

Floating platforms are generally protected by CP (sacrificial anodes or impressed current) combined with a high-quality paint system. Pipelines with an organic coating and an external concrete coating are also protected by sacrificial anodes.

For bare steel on offshore structures in seawater, the sacrificial anodes are commonly made of special aluminium alloys because these give the highest current output for a certain anode weight (Tables 8. 2 and 8. 3) as well as the lowest cost. Zn anodes are usually applied on coated and buried pipelines offshore, where the risk for passivation of Al anodes is higher due to a lower CD.

When sacrificial anodes (usually Zn or Al) are used on ship hulls they are combined with a paint system. The anodes are mounted close to each other on the area around the propeller, partly because the propeller "steals" some of the current and partly because of turbulence leading to higher demand of cathodic current density. Zn and Al anodes are also used in ballast tanks on tankers and bulk carriers.

For some other applications, such as on steel structures in soil, and in hot water tanks for fresh water, magnesium anodes are used to a higher extent. In these cases Mg anodes have a great benefit in higher driving voltage than the Al and Zn anodes. In high-resistivity environments, a combination of CP and coating is particularly beneficial with the aim to reduce the total current and thereby the potential drop near the anodes, and thus to obtain a better distribution of the current to the spots where it is needed.

For pipe systems of stainless steel carrying chlorinated seawater, internal localized corrosion can be very efficiently prevented by the application of Resistance-controlled Cathodic Protection (RCP). A resistance is simply inserted between the sacrificial anode and the pipeline, and this makes a system that is particularly suitable when there is a low diffusion-limiting cathodic current in the critical potential range. Typical of the method is that the current output from the anode is kept low, which has the consequence that the voltage drops are low and the protected pipe length from each anode is long.

Crucial properties of sacrificial anodes are listed in Tables 8. 2 and 8. 3. In aluminium, Hg (or alternatively In to avoid pollution of the environment) contributes to reduced tendency to passivation. In Zn anodes, the content of Fe (among some other species) should be kept very low to avoid self-corrosion.

Table 8.2 Typical properties of sacrificial anodes

(Design is based on somewhat higher potential and lower capacity)

Material	ρ① /Ω · cm	i_a② /A · m^{-2}	Alloying elements	Potential E_a(ref. Ag/AgCl)/V	Utilization factor u	Capacity C /A · h · kg^{-1}
Al alloys	150	1~10	Si, Fe, Zn, Hg or In, Cu	-1.05 to -1.15	0.9~0.95	2700~2830
			Si, Zn, In, Mg, Cu	-1.15	0.95	2700
			(Fe), Zn, Sn, Cu	-1.10	0.5~0.8	Varying
Zn alloys	500	1~10	Cu, Al, Si, Fe, Pb, Cd	-1.05	0.95	780
			Cu, Fe, Hg	-1.05	0.95	780
Mg alloys	6000	4~11	Cu, Al, Si, Fe, Mn, Ni, Zn, (others)	-1.50	0.5	1230
			Cu, Al or Si, (Fe), Mn, Ni, (others)	-1.70	0.5	1230

① Maximum resistirity of the water.

② Anode current density.

Table 8.3 Design data for Al- and Zn-based sacrificial anodes

Anode material	Environment	Potential E_a (ref. Ag/AgCl/seawater)/V	Capacity C /A · h · kg^{-1}	Utilization Factor u③
Al-based	Seawater	-1.05	2000①	0.75~0.90
	Sediment	-0.95		
Zn-based	Seawater	-1.00	700②	
	Sediment	-0.90		

① Anode temp. max. 25℃.

② Anode temp. max. 50℃.

③ Depends on the type of anode.

Fig. 8.13 shows various arrangements for fixing sacrificial anodes on tubular structures. Submarine pipelines are usually protected by bracelet anodes encompassing the pipe and flush with the outer concrete coating. Sacrificial anodes on ship hulls are commonly flush-mounted and welded directly to the hull.

A sacrificial anode system can be designed in the following way. The exact description of the procedure is a little different in different sources, but the principles are the same:

(1) The surface area Ac of the structure is calculated in a reasonably conservative way. Areas with different current density demand must be calculated separately.

(2) Design values of CD (I_c) on the steel surface are chosen.

(3) The total current demand, $I_t = I_c \cdot A_c$, is calculated. If the steel is coated, the current must be multiplied by a coating breakdown factor f_b.

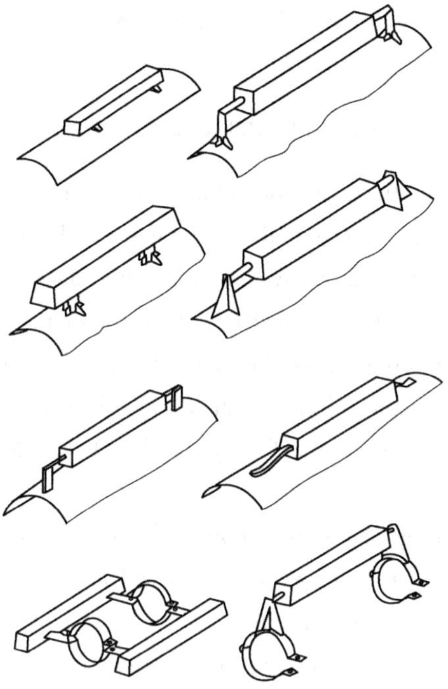

Fig. 8.13 Typical arrangements for fixing sacrificial anodes on offshore tubular structures

(4) Material, shape and size of the anodes are selected, and the anode resistance R_a is calculated by use of the appropriate formula.

(5) The current output from each anode is given by

$$I_a = \frac{\Delta P_a}{R_a} \tag{8.4}$$

where P_a is the potential difference that is available for overcoming R_a.

(6) The number of anodes is calculated:

$$n \geqslant \frac{I_t}{I_a} \tag{8.5}$$

It should be noticed that both the current demand I_t and the output from each anode I_a is reduced during the service time. The number of anodes should therefore be calculated both for the initial and for the final stage. The largest of the calculated n-values is chosen.

(7) The total weight of anodes necessary to protect the structure during the specified lifetime t is calculated:

$$G_t = \frac{I_g t}{Cu} \tag{8.6}$$

where I_g is the average (mean) current demand $= \sum i_m A_c$, C is the anode capacity and u the utilization factor (Tables 8.2 and 8.3). Then it is checked that

$$n \geqslant \frac{G_t}{G_a} \tag{8.7}$$

where G_a is the weight of each anode. E_a, I_a, C, u and other anode data are available from anode suppliers.

8.3.1.4 Cathodic Protection with Impressed Current

When such a system is used on fixed offshore platforms it may be preferable to combine it with sacrificial anodes (hybrid system). Then, the sacrificial anodes can protect the structure before the impressed current installation is ready for use, and also contribute to a better current distribution. The largest benefits of impressed current compared with sacrificial anodes are lower anode weight and lower drag forces from the sea. Also, relatively few anodes are necessary. Theoretically, it is a more economic solution than sacrificial anodes, but practical experience with cases of serious mechanical damage of the CP system has led to fewer applications of impressed current systems.

On large ship hulls and floating platforms, impressed current has often been used in combination with a paint coating. The same is the case for land-based buried structures.

Anodes of a corrosion-resistant material such as Pt, Pb, Sb, Ag, graphite, magnetite or high-silicon iron are normally used in impressed current installations. Pt is often used as a thin layer on a substrate of another material, e.g. in the form of platinized titanium. A corroding material, such as scrap steel, can also be used, but additional anode material must be supplied regularly in this case.

The anodes can be arranged in different ways. On fixed offshore structures the anodes may be mounted on the structure itself (but of course insulated from it), they may be supported by special conduits more or less remote from the structure, or remote seabed anodes may be used.

On ship hulls protected by impressed current, flush-mounted anodes are used in order to avoid additional hydrodynamic resistance. Around the anode, the nearest steel surface is covered with a dielectric shield or coating with the aim of obtaining a better current distribution.

A land-based arrangement for CP in soil is shown in Fig. 8.14. A backfill of a carbonaceous material such as coke breeze or graphite particles is used around the anode. The backfill is electron-conducting, which makes the effective size of the anode larger. This in turn reduces the resistance between the anode and the structure, and it reduces the consumption of anode material. If the resistivity of the soil is very high, buried pipelines may be protected by means of a continuous anode along the pipeline at a suitable distance from it.

8.3.2 Anodic Protection

Anodic protection can be applied on materials with a well-defined and reliable passive region and low passive current density, i.e. with an anodic overvoltage curve like that shown in Fig. 8.15. The material is polarized in the anodic direction so that the potential is lifted to the passive region. As indicated in the figure, the current density necessary to hold the potential E_a at a suitable value is low, and this makes the method economic in use. However, the initial current density i_{emax} may be relatively high. The low current density in the established protection state leads to a high

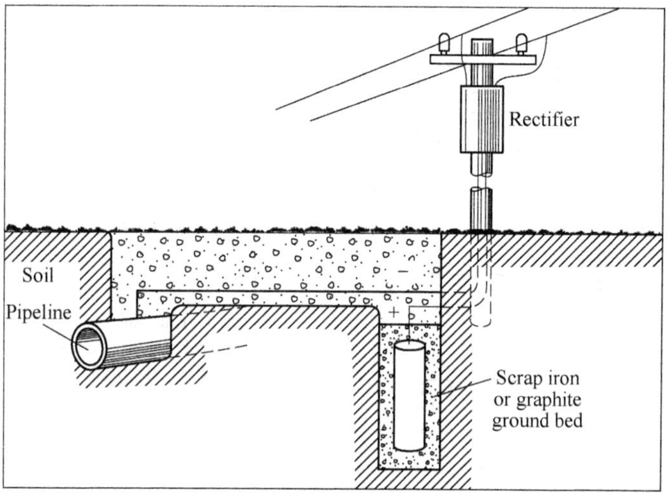

Fig. 8. 14 Cathodic protection of a buried pipeline by impressed current

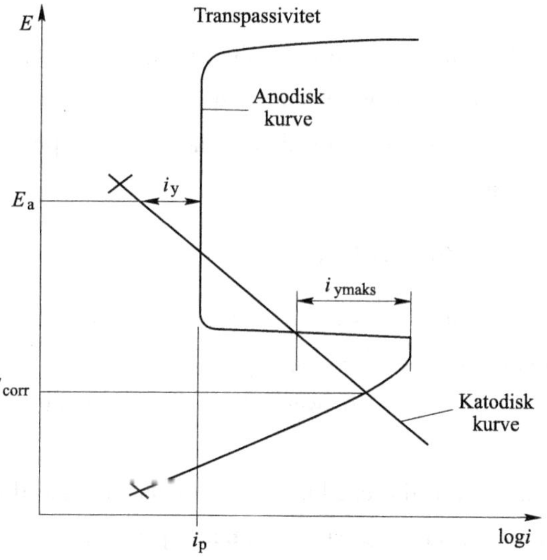

Fig. 8. 15 Potential shift from the corrosion potential E_{corr} to a higher potential E_a by means of an external current with a final value $I_e = i_e A$

throwing power of the current, i. e. areas "in shadow", internal corners etc. are adequately covered, keeping the whole surface of the component in the passive state. Fig. 8. 16 shows in principle the arrangement for internal protection of a steel tank by means of a potentiostat.

Anodic protection is used on objects such as steel tanks for storing and transport of sulphuric acid, apparatus made of stainless steels and titanium for treatment of various acids and salt solutions, and for aluminium exposed to water at high temperature. The method cannot be used in aggressive liquids that may cause localized corrosion or high passive current density. It may be used on 18/8 CrNi stainless steel exposed to 30% H_2SO_4+ 1% NaCl and on titanium exposed to hydrochloric acid solutions. Pitting is avoided, in the former case because the sulphate ions counteract the chloride, in the latter case because the pitting potentials of titanium in the actual solutions are very high.

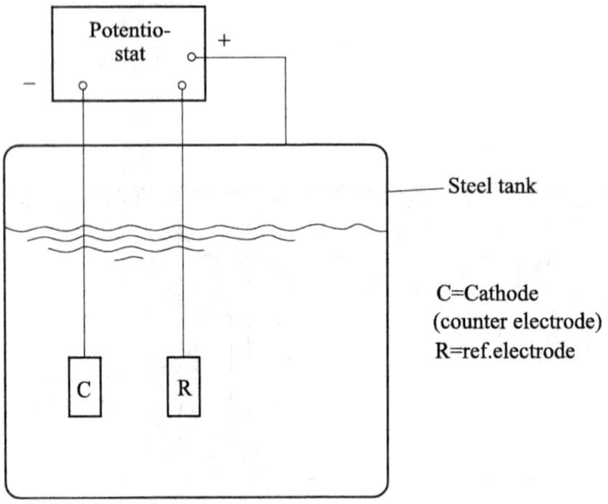

Fig. 8. 16 Internal anodic protection of a steel tank

Anodic protection might be utilized much more than it has been so far, but the method must not be used under unfavourable conditions, because the anodic polarization may cause a strong increase of the corrosion rate. If there is any doubt, laboratory experiments may clarify if anodic protection can be recommended.

8. 3. 3 Corrosion Protection by Coatings

Through the application of coatings, corrosion is prevented by one of the following three main mechanisms or by combination of two of them:

(1) Barrier effect, where any contact between the corrosive medium and the metallic material is prevented.

(2) Cathodic protection, where the coating material acts as a sacrificial anode.

(3) Inhibition/passivation, including cases of anodic protection.

In the following sections, a survey of different types of coating is given, which includes properties, mode of action, fields of application, and coating methods. Without regard to the type of coating and the application method, a proper pre-treatment of the material to be protected is necessary for obtaining a good result.

8. 3. 3. 1 Metallic Coatings

In most cases of corrosion protection by *metallic coatings*, the purpose is to protect unalloyed or low-alloy steel, but there also exist many cases of other metals to be protected this way. Metallic coatings can be divided in two groups, the cathodic coatings, which are nobler than the substrate, and the anodic ones, which are less noble than the substrate, i. e. the coatings that have, respectively, a higher and a lower corrosion potential than the substrate in the environment in question. The cathodic coatings will most often act by the barrier effect only, but for some combinations of substrate and environment the substrate can also be anodically protected (on uncovered spots). The anodic coatings will in addition to the barrier effect provide cathodic protection of possible "holidays", i. e. spots or parts of the surface where the coating is imperfect and the substrate is exposed to the corrosive environment. The normal major difference between a cathodic and an anodic coating is just the behaviour at such a defect. This is illustrated in Fig. 8. 17. In the case of a cathodic coating (a) the substrate is subject to galvanic corrosion in the coating defect. The corrosion may be rather intensive because the area ratio between the cathodic coating and the anodic spot of bare substrate usually is very high. In the other case (b), only a cathodic reaction occurs on the bare substrate (it is protected cathodically), while the coating is subject to a corresponding galvanic corrosion distributed over a larger surface area. In order to protect the substrate, low porosity, high mechanical strength and continuous adhesion are even more necessary for a cathodic than for an anodic coating. In relation to steel, Ag, Ni, Cr and Pb are cathodic, while Zn and Cd are anodic in most environments. The polarity of Sn and Al referred to steel varies from one environment to another.

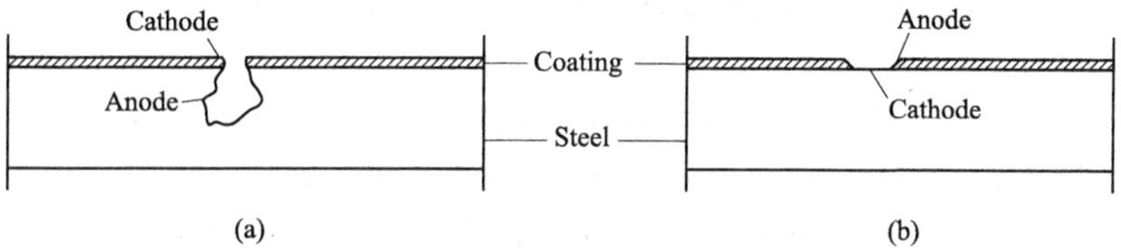

Fig. 8. 17 Localization of corrosion at a defect in a metal coating on steel
(a) Cathodic coating; (b) Anodic coating

Silver is used as a coating material for decoration purposes; furthermore its chemical and electrical properties are utilized, respectively, in various equipment and environments in the process industry and in electric contacts and other small components in electrical and electronic equipment. Silver coatings are commonly applied by electrolytic plating. A coating of copper or nickel is often used under the silver top coat. Silver may be attacked by sulphides in industrial atmospheres. To prevent attack, sometimes a thin layer of rhodium is used on top of the silver coating; alternatively, a passivating chemical treatment is carried out.

Nickel has often been used for preventing corrosion and for providing a nice surface on steel parts in contact with ordinary atmosphere, e. g. bicycle parts, car parts and fittings. In order to make the surface brighter and more corrosion resistant the nickel plate is covered by a thin coating of chromium. The chromium coating has a large number of micropores or microcracks, which is advantageous because they make the corrosion of the underlying nickel well distributed and shallow. The chromium layer is usually deposited on top of a thin layer of bright nickel, which is anodic in relation to a thicker basic layer of (semi) matt nickel. Because of its less noble nature, the bright nickel layer gives a certain cathodic protection effect on the basic layer in existing pores.

A basic layer of copper may be used on iron and steel substrates, which allows a thinner nickel coating. Nickel-chromium coatings can also be used on zinc (with a basic layer of copper), on copper alloys and aluminium alloys.

For industrial use, nickel coatings are applied without chromium on top, and the coatings are most often thicker, usually 25 ~ 250μm, and up to 500μm for some applications. Such coatings are most suitable for environments that tend to cause general corrosion with reaction products of low solubility (environments without chlorides). However, it is also stated that nickel performs well in seawater provided that the flow rate is high. Furthermore, the experience with nickel in various alkalis, neutral and alkaline salts and organic chemicals is good.

Electrochemical deposition (electroplating) has traditionally been, and is still, the dominating technique for application of nickel coatings. In recent years, the number of applications of chemical (electroless) deposits of nickel has been growing (parts in pumps, valves, compressors, storage vessels for chemicals etc).

Nickel coatings can increase the resistance to corrosion fatigue, but stresses set up by the coating may reduce the pure mechanical fatigue strength of the base material. The effect can be counteracted by compression stresses in the coating, which can be obtained with certain electrolytic baths, or by applying a soft basic coating (tin, lead), which can prevent cracks in the coating propagating into the base material.

Thick or so-called hard chromium coatings are used directly on steel, cast iron and light alloys to improve resistance to wear and corrosion in hydraulic cylinders, pump shafts, bearings, dies etc. Thermodynamically, chromium is less noble than iron, but under atmospheric and many other conditions it is practically more noble due to the strong tendency to passivation.

Lead is particularly suitable in coatings exposed to dilute sulphuric acid and industrial atmospheres.

Tin shows rather complicated corrosion properties in contact with steel. In the atmosphere and in solutions of natural inorganic salts, Sn is cathodic in relation to Fe. In contact with fruit juice, meat, milk and alkaline solutions, tin are anodic. Tin plating is most often carried out by electrochemical deposition, but hot dipping is also common.

The Pourbaix diagram for zinc is shown in Fig. 8. 18. The boundaries of the passivity region shown by solid lines are valid for water without CO_2. For pH<8 ~ 9, the following reactions occur:

Anodic: $$Zn \longrightarrow Zn^{2+} + 2e^-$$

Cathodic: $$2H^+ + 2e^- \longrightarrow H_2$$

$$O_2 + 2H_2O + 4e^- \longrightarrow 4OH^-$$

The equilibrium potentials of the three reactions as functions of pH are drawn in the diagram. For pH values between 8. 5 and 11, zinc is passivated by formation of $Zn(OH)_2$. When CO_2 is dissolved in the corrosive medium the passive region is enlarged so that it reaches from pH = 6 to pH = 11 (dashed lines). The reason for this is the formation of a protective film of alkaline zinc carbonate. The effect of this film is strong in rural atmospheres, in mildly aggressive city atmospheres and in neutral and slightly acidic rain water. Also the calcium content in hard waters leads to deposits that prevent corrosion. Conversely, in various salt solutions the formation of soluble corrosion products may counteract the passivation process. Chlorides act in this way, but in marine atmospheres and seawater, the detrimental effect of chloride is at least partly balanced

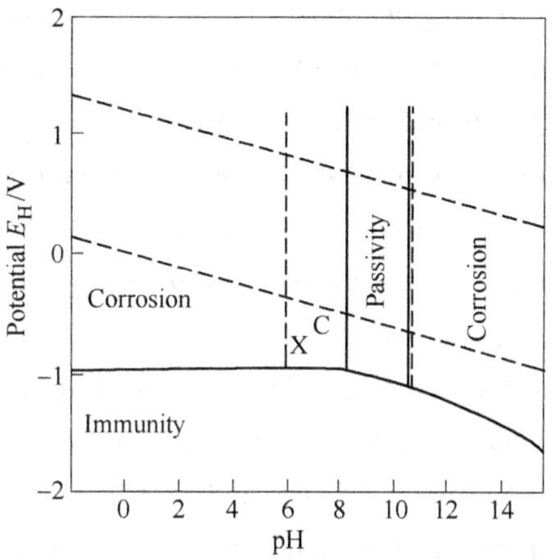

Fig. 8.18 Pourbaix diagram for Zn in water at 25℃

by protective calcareous deposits. Corrosion of zinc in seawater is therefore mainly under cathodic control, which is also indicated by the level of the corrosion potential (marked C in the diagram in Fig. 8.18). In industrial atmospheres with high SO_2 content, zinc sulphate and zinc sulphite are formed, which are easily washed off by rain. Under such conditions, the corrosion rate will be relatively high, in agreement with the indication given by the Pourbaix diagram for acid environments.

Another important factor is the temperature. The corrosion rate at 70℃ may be 50 times that at 20℃. Above 60~70℃, zinc is cathodic in relation to iron and carbon steel (see below).

Zinc is easy to apply electrochemically, by hot dipping, and by thermal spraying. Electrochemically applied coatings are relatively thin, usually up to 25μm, and they are in most cases post-treated with a converting process, such as chromating, which gives a yellow, blue, drab or colourless appearance. Hot dipped zinc coatings are often 80~100μm thick, and a typical thickness range for thermal spray Zn is 100~200μm, and these two methods can therefore provide more long-term protection under suitable conditions.

Contact between zinc and steel. In spite of its electronegative character, zinc corrodes relatively slowly when separated from other metallic materials. The reasons are that the reaction $2H^+ + 2e^- \rightarrow H_2$ does not go fast on Zn, and that (in the atmosphere) protective carbonate is formed. As a coating on a more electropositive material such as carbon steel (at <60℃), the corrosion rate of zinc is increased near the bare steel regions. The potential at these places will increase to some extent compared with the normal potential on zinc, but it will still remain below a practical limit for corrosion of steel. For larger uncovered regions, the following should be noted: with increasing distance from the zinc, the potential of the steel surface increases, and at a certain distance significant corrosion of steel will occur. This critical distance increases with increasing conductivity of the liquid and its corrosivity versus zinc. However, at the same time as an increased area of bare steel can be protected, the corrosion rate of zinc may be so high that protection of steel with zinc is not an economic method. In such cases, coating the steel with aluminium may be a better alternative, but possible bare steel regions must then be kept small.

On the basis of the electrochemical nature of corrosion, it is clear that metallic zinc must be in direct contact with the electrolyte to protect bare steel regions. Paint or other coatings on top of the zinc coating will therefore eliminate or at best reduce the cathodic protection effect.

The change of polarity between the steel substrate and zinc coating when the temperature is increased to 60 ~ 70℃ may cause corrosion of steel on bare regions. Therefore, a continuous coating is required at higher temperatures.

Aluminium. As a coating material it is applied in thermal spraying (metal spraying) both on land-based and on offshore steel structures. Hot dipping is a more sensitive process with aluminium than with zinc as coating material, and it has not obtained the same widespread application for Al as it has for Zn. Hot-dipped coatings produced of mixtures of aluminium and zinc have shown good all-round properties for atmospheric exposure. Aluminium can also be plated by electrochemical as well as chemical deposition. Cladding of surfaces can be carried out by rolling, extrusion, drawing, and explosive bonding.

Aluminium in contact with structural steel. Although steel corrodes faster than aluminium in environments such as seawater, marine and industrial atmospheres, and water containing SO_2, and soft water, aluminium is the material that primarily corrodes when these two materials are galvanically coupled in the mentioned environments. The reason for this apparent paradox is that the corrosion potential of steel is higher than on aluminium when the materials are separated. By the coupling a mixed potential is established which is lower than the separate steel potential and above the corrosion potential of aluminium. Particularly in seawater, the corrosion of aluminium is accelerated considerably by the galvanic contact. But the relationship may be more complicated in some other environments; aluminium may in one case be anodic and in another cathodic to steel. In pores in thermally sprayed aluminium on steel in moist atmospheres, the following phenomena may occur successively:

(1) The coating acts as a cathode due to the oxide, which prevents conductance of ions but allows electronic conductance to some extent.

(2) Due to increased pH at the cathode, the solubility of the aluminium oxide increases. Therefore, metallic aluminium is exposed to the solution, the coating becomes less noble than steel and begins to act as an anode.

(3) The corrosion products of Al gradually block the pores in the coating, the galvanic element is made ineffective and the coating becomes stable.

Comparison between zinc and aluminium as coatings on steel. Due to the oxide, aluminium has not the same remote effect as zinc for protection of a bare steel area. In acid solutions (pH = 4~5) and severe industrial atmospheres the consumption of zinc may, however, be so large that an Al coating is a more economic solution. It has been claimed that the lifetime of aluminium in industrial atmosphere may be three times that of a Zn coating. Also exposed to soft water, aluminium is considered superior to zinc. Zinc is preferable in hard water and alkaline solutions, but for two different reasons. In hard chloride-free water, Al corrodes too slowly to be able to protect the steel in coating defects. Conversely, in alkaline solutions aluminium corrodes too fast, which results in too short a lifetime of the coating. Aluminium normally corrodes more slowly than zinc in marine atmospheres and seawater. The remote effect of aluminium is satisfactory in seawater under most conditions. Unfavourable localized corrosion effects may occur on a sprayed aluminium coating if it is galvanically coupled to a large bare steel surface.

Coatings of Zn and Al can be combined with an external paint coating, with good results when the work is done properly. An alternative is to use a thin external organic sealer, which blocks the pores in the metal coating. Because Al and Zn coatings are relatively inefficient as cathodes, they can also be combined with cathodic protection, particularly on structures that are going to have a long life in the sea. Aluminium is favourable with regard to current density demand and potential. In some cases, thermal spraying of aluminium has therefore been applied on submerged parts of offshore steel structures. Compared with bare steel, this protection system gives a strong reduction of the cathodic current demand.

In the selection of a coating method, there are usually more real possibilities with zinc than with aluminium. For steel structures, aluminium has mainly been competitive on price only in the form of thermal spray coatings.

Cadmium has been a very useful coating material, but because of its toxic nature the numbers of applications has been reduced during recent decades.

8.3.3.2 Other Inorganic Coatings

Coatings in this group prevent corrosion by their barrier effect. The group comprises, among others, various types of enamel and glass linings. The coating material, which must have a suitable thermal expansion coefficient, is applied in the form of a powder (after pickling or other cleaning of the base material) and heated in a furnace until the lining material becomes soft and is bound to the metal surface. High resistance to numerous chemicals and other environments can be obtained by variation of the composition of the coating material. The linings are brittle and can tolerate very little deformation, and some of them are sensitive to thermal shock.

Coatings of ordinary mortar (from Portland cement) have nearly the same thermal expansion coefficient as steel, and they are cheap and easy to repair. They can be applied with a thickness in the range 0.5~2.5cm and more to protect pipelines and tanks, internally or externally. Thicker coatings are often reinforced with a net. On internal cylindrical surfaces the coatings can be applied by centrifugal casting, or by spraying and manual finishing. Imperfections and damaged parts in the coating may cause concentrated attacks, particularly if there is a galvanic contact with more noble materials than steel, which can be the case in pipe systems including valves, pumps etc. The problem is avoided if the coating is combined with cathodic protection.

Ceramic coatings of oxides, nitrides, borides etc. can be applied by thermal spraying. Some of the ceramic materials can be sprayed by the classical flame spraying method, while others must be plasma sprayed because the melting points are too high for the former method. Ceramic coatings are particularly used to prevent high-temperature corrosion and wear and to provide thermal barriers (thermal insulation). Important alternatives in many cases are ceramic-metallic coatings that most often are composed of hard carbide particles bound together by a metallic material. These materials are also called hardmetals. A common composition is WC with a Co binding phase. However, the corrosion properties of cobalt are rather poor, and it is preferably replaced by an alloy of Co and Cr or an alloy of Ni, Cr and possibly some other elements for application under corrosive conditions. These coatings are among the best at resisting abrasion and erosion possibly combined with corrosion. With modern high-velocity flame spraying (HVOF) denser and better coatings are obtained.

Conversion coatings. This group comprises phosphate, chromate and oxide coatings. Phosphate treatment of steel is usually carried out by exposing the steel surface to phosphoric acid containing phosphates of Zn and Mn, either by spraying or by dipping after prior acid pickling or blast cleaning. Phosphate coatings that are only 1 ~ 10μm thick do not provide a complete protection by themselves, and the process is mainly used as pre-treatment before painting or treatment with wax or oil. Metals such as Zn, Cd, Al and Mg can also be phosphated. Chromate treatment has been widely used on non-ferrous metals, particularly Al and Zn, sometimes as the only treatment and sometimes as pre-treatment before painting. Since chromates are highly toxic the environmental regulations connected with chromate processes have become more stringent, and several efforts have been made to find more environment-friendly conversion alternatives. Oxidation of steel by dipping in hot concentrated alkaline solutions containing persulphates, nitrates and chlorates is used as a base for oil and wax. An important process is oxidation of aluminium and aluminium alloys. It is achieved by anodic polarization, and is usually called anodizing. The oxide film can be dyed in the anodizing process by various additions to the bath, or it can be dyed afterwards. It is sealed by treatment in boiling water, steam or possibly in a hot, dilute chromate solution. Anodizing provides in itself some improvement of the corrosion resistance of aluminium, and furthermore it gives a good base for painting.

For example, a major barrier restricting the applications of BP is its successive decomposition upon long-term exposure in moisture atmosphere. The underlying mechanism for BP self-degradation is mainly due to the formation of P_xO_y species via the reaction between oxygen and the lone pair electrons perpendicular to the BP surfaces. To mitigate this problem as well as regulate the performance of phosphorene, various strategies like covalent, vdW and electrostatic functionalization have been devised. Herein, covalent modification is controlled by bonding where the lone pairs of BP surfaces interact with Lewis acids such as transition metal ions, organic molecules and alkali ions. vdW functionalization features the interactions between BP and a capping layer like graphene, hBN, and organic molecules. As for electrostatic functionalization, positively charged adsorbates may spontaneously interact with phosphorene modified via negatively charged species. Ideally, functionalization can enhance the stability of phosphorene while at the same time retain or promote its specificity for long-term applications. In our study, we constructed a 0D-2D inorganic-organic composite combined ultrasmall BPQDs with ultrathin g-C_3N_4 nanosheets via a conventional and cost-effective sonication approach. BPQDs with an average size of 4. 2nm were decorated onto the surface of g-C_3N_4 nanosheets. As compared to pristine BPQDs, the nanohybrid with an optimal BPOD content ($w_{BPQDs/g\text{-}C_3N_4}$ = 5%) presented the improved photoreactivity towards water splitting under visible-light ($\lambda \geq 420$nm) illumination. Specifically, pristine BPQDs only show a H_2 production rate of 65μmol/(h · g) in the initial 1.5h and a negligible amount of H_2 is detected afterwards. Under the identical test conditions of durability, no significant photoactivity loss for BPQDs/g-C_3N_4($w_{BPQDs/g\text{-}C_3N_4}$=5%) is observed after 8 cycles up to 32h. After long time photoreaction, the structure and compositions of $w_{BPQDs/g\text{-}C_3N_4}$ hybrids are maintained. Our results verified that the integration of BPQDs with g-C_3N_4 nanosheets is a promising means to construct an efficient 0D-2D hybrid photocatalyst for the direct conversion of sunlight to renewable solar fuels. Aberration-corrected scanning transition electron microscopy (AC-STEM) in conjunction with Fourier transform infrared spectroscopy (FTIR) characterization confirmed that phosphorus-carbon (P-C) bond are formed at the interface between BPQDs and g-C_3N_4, which could not only stabilize the structure of the prepared photocatalytic system, but also prevent the further oxidation of BP in ambient conditions and then prolong the lifetime of the obtained photocatalysts. Also, we concluded that the type-Ⅱ band alignment, the formation of phosphorus-carbon bonds and the associated efficient spatial separation of photoinduced charge carriers across the interface between BPQDs and g-C_3N_4 were crucial for the improved performance.

8.3.3.3 Paint Coatings

The use of *paint coatings* is the most common method for corrosion prevention.

Compositions and types of paint. An anticorrosive paint is composed of a binder, pigments, a solvent/diluent, extenders and a variable number of other additives such as antioxidants, surface-active agents, driers, thickeners and antisettling agents. A paint is primarily characterized by its pigment or by its binder. We distinguish between primers, which usually contain pigments causing some inhibition or cathodic protection of the substrate, and paints for finishing coats, which contain colour pigments and extenders, which may improve the barrier effect of the coating system. The coats in thicker paint systems may be divided into primer, intermediate or body coats and topcoat. Recent development of thick film paints implies less need for distinction between intermediate coats and topcoat.

An actual inhibiting type of pigment in primers is zinc phosphate, while red lead and zinc chromate, which were earlier in widespread use, are seldom applied nowadays because of the risk of health injuries. Metallic zinc powders provide cathodic protection of the substrate if the zinc concentration is high enough. The pigments in the finishing coats provide colour and protect the binder from being damaged by ultraviolet sunlight. The most efficient barrier-acting pigment consists of aluminium flakes. The aluminium flakes increase the length of the diffusion path and are therefore assumed to increase the resistance to diffusion of water, oxygen and ions, but more recent investigations have shown that the effect and the mechanism depend strongly upon the type of binder. Other commonly used pigments in intermediate coats and topcoats are various forms of iron oxide.

The binder may, for instance be bitumen (coal tar or asphalt), or linseed oil (natural materials), alkyd, chlorinated rubber, epoxy, vinyl, polyurethane etc. (synthetic organic materials), or silicate (inorganic). Some paints are hardened by reaction with oxygen in the air (oil and alkyd paints), others by evaporation of the solvent (e. g. chlorinated rubber) and a third group by a chemical reaction between two components (epoxy and polyurethane paints).

The primary questions that have to be asked before selecting a paint system are:

(1) Which pre-treatment is possible, and what will the condition of the substrate be before painting?

(2) How will the environment around the painted structure change during various periods of the lifetime? Which mechanical and chemical attacks will the coating be exposed to?

(3) What are the conditions for application and drying/hardening of the paint, particularly the temperature and the humidity?

(4) What are the initial and maintenance costs of paint and painting work?

Several so-called rust-stabilizing or rust-converting products are commercially available, but the practical efficiency of these products depends on the conditions. They are made for coating on rusty substrates, but it is still important to remove loose rust, which can be done by thorough steel brushing. Some products are based on phosphoric acid. These must be applied in appropriate thickness, and excessive acid must be washed away with fresh water before the coating is overpainted.

A zinc-rich primer is often recommended. It can be an organic zinc-epoxy or an inorganic zinc-ethyl-silicate primer. Zinc-rich primers are also used as so-called shop primers, or prefabrication primers, for temporary protection of semi-manufactured steel goods. After fabrication, e. g. of welded steel structures, the shop primer surface must be cleaned (degreased), and possible shop primer defects and weld joints have to be blast cleaned and coated with a primer before the whole structure is painted. Iron oxide is also used as a pigment in some shop primers. These must not be overpainted with a zinc-rich paint.

The main rule for painting previously painted steel is to use the same type of paint. If the old paint coating has turned out to be unsuitable in the actual environment, it should be removed before the structure is coated with a different type of paint.

The cost of the paint itself forms often 15% ~ 20% of the total cost of the painting operations including pre-treatment and application. It is important that the pre-treatment and the type of paint are compatible. The advanced paints depend upon good pre-treatment to obtain the necessary adhesion to the substrate.

The rate of chemical reactions normally decreases with decreasing temperature. Therefore, two-component paints and oxidative hardening paints require an application temperature above a certain minimum. The steel temperature is more important than the air temperature. Temperature limits often used are: epoxy paints: 10℃, polyurethane: 0℃, oxidative hardening paints: 0℃. If it is strictly necessary to paint in cold weather, physically drying paints (i. e. those drying by evaporation of the solvent) should preferably be used, since these dry relatively fast also at low temperatures.

Moisture is detrimental to the application of paint, and condensation may sometimes be a problem. All paints give the best result when they are applied on a clean and dry substrate. However, special paints based on an alcohol solvent are more tolerant than others versus moisture. Furthermore, zinc-ethyl-silicate has to absorb water from the air in order to dry, and in this case the relative humidity of the air should not be too low either. Conversely, pure vinyl paints are particularly sensitive to high humidity. Painting a moist surface should be done by brush rather than by spraying. Special paints that can be applied and that harden submerged in water have also been produced.

A suitable film thickness and appropriate periods between applications of the successive coats are important, but depend upon the type of paint. Data sheets from the paint producers give information about this. The thickness should be checked during the painting work. Other important properties are adhesion between old and new paint as well as resistance to detergents and mechanical wear. It can be mentioned that hardened two-component paints may be less suitable for over painting unless they are rubbed mechanically, but on the other hand they are resistant to detergents and mechanical wear.

Various thermal, mechanical, chemical and biological conditions can cause degradation of coatings. UV radiation is an important factor in the atmosphere. Corrosion underneath the coatings is also a common cause of paint degradation, which may result in phenomena such as blistering, anodic undermining and cathodic disbonding. Anodic undermining means that the coating is undermined by the anodic dissolution of the substrate, often at the edge of a defect in the coating. Cathodic disbonding implies that the coating loses its adhesion to the substrate because of the cathodic reaction. The formation of a local alkaline environment is a crucial step in the process. The cathodic reaction may be part of a corrosion process, or a result of cathodic protection. The attack starts at a weak point in the coating and develops radially, most often forming a circular defect. An interesting example of this is shown in Fig. 8.19, where we can see a few such circular defects that have been overpainted. At one of them (in the centre) a new defect has developed from the boundary of the overpainted defect.

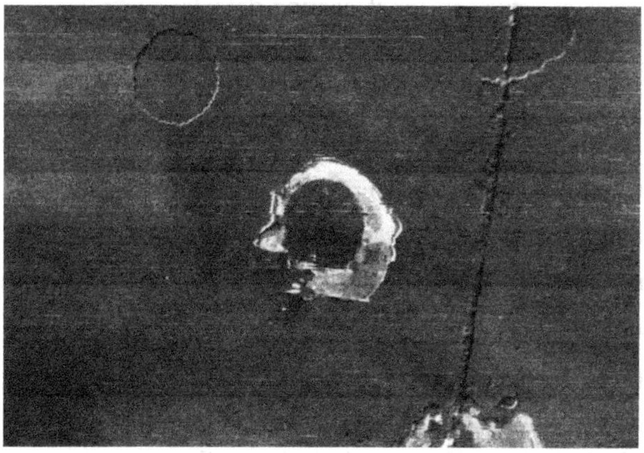

Fig. 8.19 Cathodic disbonding of a paint coating on a cathodically protected ship hull
(The defects developed during the first period have been overpainted)

Painting zinc and aluminium makes special demands on the pre-treatment of the substrate. Chromate treatment provides a good basis for paint, and it is still in widespread use in spite of the toxicity of chromate. Phosphate treatment or wash-priming can also be used. Alternatives for aluminium are light blast cleaning or anodizing.

By properly performed painting work we can avoid defects such as "holidays" (areas left uncoated), flaking (e.g. due to inadequate pre-treatment of the substrate), too thin or too thick coats (insufficient checking of thickness) and blistering (which may, for instance occur because of short hardening periods between application of successive coats combined with low temperature). In Fig. 8.19, an example of blistering in a chlorinated rubber coating is shown. The blistering has in this case been accelerated heavily due to heating by sunshine after completed painting. Regarding coating thickness, particular attention should be paid to edges, welds and other irregularities on the base material. Thick-film paints give better coverage on such places.

8.3.3.4 Other Forms of Organic Coatings

Rubber coatings provide excellent corrosion protection in seawater and several chemicals, and are used internally in tanks and pipelines. Externally on buried pipelines, asphalt or coal tar has commonly been used, often reinforced with a textile or mineral net, and during the last generation, also tape made from materials such as polyvinyl chloride or polyethylene with adhesive and primer on one side. Various factory methods for direct application of polyethylene or powder epoxy coatings on pipelines have been developed and used. During the last few decades, glass-flake-filled unsaturated polyester or vinyl-ester coatings have been applied to an increasing extent, particularly on offshore installations, on ships and in industry. These are fast-hardening coatings with high mechanical wear resistance and with good ability to protect against corrosion in most environments. The coatings are applied by high-pressure spraying in 1 ~ 2 coats giving a thickness of 0.75 ~ 1.5mm.

Depending on the environment, all the mentioned types of coating can be combined with cathodic protection to prevent corrosion in/at coating defects.

8.3.4 Change of Environment

The environment may be changed in the following ways in order to reduce corrosion rates:

(1) Decreasing (or increasing) the temperature;

(2) Decreasing (or increasing) the flow velocity;

(3) Decreasing (or increasing) the content of oxygen or aggressive species;

(4) Adding inhibitors.

On the following pages, various groups of inhibitors, their effects, and examples of compositions and applications are presented. The inhibitors can be arranged in groups based on which reaction (anodic or cathodic) they affect and how they influence upon the polarization properties. Five different groups are characterized by the overvoltage curves shown in Fig. 8.20(a) ~ (e).

The passivating inhibitors, also called passivators ((a) and (b) in Fig. 8.20), are usually inorganic. The oxidizing ones (a) act by depolarizing the cathodic reaction (making it more efficient), or more frequently by introducing an additional cathodic reaction. When the concentration of inhibitor becomes high enough (higher than a critical value c, see Fig. 8.20(a)), the cathodic current density at the primary passivation potential becomes higher than the critical anodic current density, and consequently the metal is passivated.

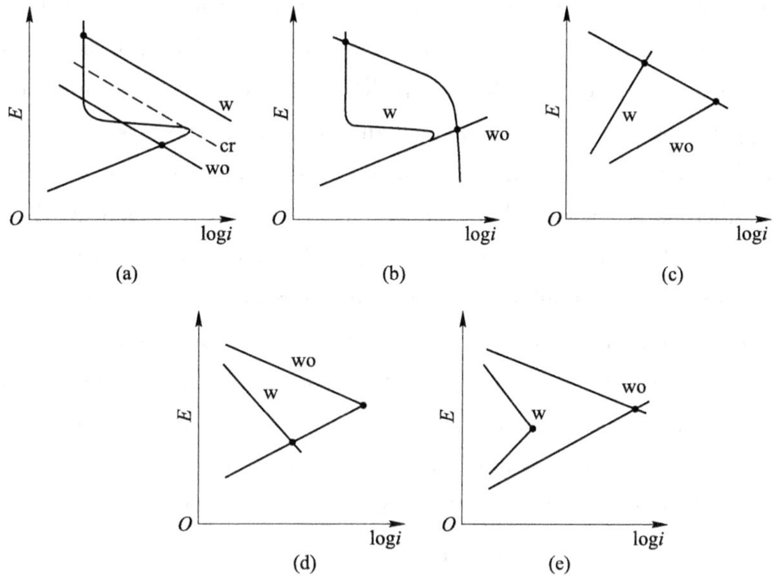

Fig. 8.20 Effect of different types of inhibitor on cathodic and anodic overvoltage curves (w = with and wo = without inhibitor. cr = critical concentration of inhibitor)

(a) Oxidizing inhibitor (passivator); (b) Anodic, passivating inhibitor (passivator); (c) Anodic, not passivating; (d) Cathodic; (e) Double acting inhibitor

Examples of oxidizing inhibitors are chromates and nitrites. Critical concentrations of these for steel in neutral water are $10^{-4} \sim 10^{-3}$ M at room temperature. At 70 ~ 90℃ as much as 10^{-2} M is needed because the critical current density increases by increasing temperature. Chlorides in the water affect the critical concentration in the same direction as does the temperature. Chlorides and sulphates also lead to higher passive current density and reduce the efficiency of the inhibitor. Chromates and nitrite inhibitors may with advantage be combined with pH adjustment to values between 7.5 and 9.5, at which the critical concentration of inhibitor is lower than at lower pH values.

Both chromates and nitrites are often used in cooling systems. Chromates are poisonous, which is a serious drawback. Nitrites are also advantageous in cooling water with antifreeze liquid on an alcohol or ethylene glycol basis since they, contrary to chromates, do not react with the antifreeze liquid. Furthermore, nitrites are used in oil-water emulsions in machining, in oil storage tanks in houses and internally in gas and oil-carrying pipelines. A common nitrite inhibitor is $NaNO_2$.

When using oxidizing inhibitors it is very important that the inhibitor concentration is high enough everywhere in the system because a concentration below the critical value is worse than no inhibitor at all. At positions with stagnant liquid there may be a risk of lower supply than consumption of inhibitor and consequently too low a concentration, resulting in localized corrosion. In order to avoid this, the liquid should be kept moving, and crevices as well as deposits should be avoided when using such inhibitors. It is also important to check the inhibitor concentration regularly.

The described effect of the oxidizing inhibitors implies that they are primarily suitable for active-passive materials, steel being the most common example. However, it has been demonstrated empirically and experimentally that, e. g. sodium chromate is efficient also on aluminium in water-containing chlorides, some acids and slightly alkaline solutions, and that chromates are useful for protection of copper alloys, soft solder joints and zinc in cooling water systems. This indicates that chromates also affect the anodic reaction, i. e. partly as an anodic passivating inhibitor. Generally, this group can rather be characterized by the overvoltage curves in Fig. 8. 20 (b). It comprises alkaline compounds such as NaOH, Na_2PO_4, Na_2O_n, SiO_2 and $Na_2B_4O_n$. Contrary to the oxidizing inhibitors, the anodic passivating inhibitors usually lead to reduced corrosion rate only in waters with dissolved oxygen. Within this group we have examples such as sodium benzoate and sodium polyphosphate. The latter may in some systems act by forming diffusion-limiting layers of deposits on cathodic areas (similar to some cathodic inhibitors, see below). The mentioned anodic passivating inhibitors are particularly suitable for iron and steel in near neutral waters.

Non-passivating inhibitors include both some anodic (Fig. 8. 20 (c)) and the cathodic ones (Fig. 8. 20(d)). The latter are exemplified by those that remove free oxygen by a reaction, as is the case for hydrazine and sodium sulphite:

$$N_2H_4 + O_2 \longrightarrow N_2 + 2H_2O \tag{8.8}$$

$$2Na_2SO_3 + O_2 \longrightarrow 2Na_2SO_4 \tag{8.9}$$

These inhibitors are effective in all environments where the oxygen reduction is the dominating cathodic reaction in the uninhibited state, such as in neutral natural waters. Conversely, other types must be used in strongly acid solutions. Typical inhibitors for strong acids are the ones used as pickling inhibitors (in pickling for removal of mill scale or for cleaning metal surfaces in other connections). Pickling inhibitors are most often organic substances that are adsorbed on the surface and are double acting, in the sense that both the cathodic and the anodic reaction are made less efficient (Fig. 8. 20 (e)). They are often added in concentrations of 0. 01% ~ 0. 1%. Organic, adsorption species form the largest group of inhibitors. Typical examples in this group are organic amines. The group includes—among others—substances that are used in oil, grease or wax on machine parts during transport and storage.

Vapour phase inhibitors can also be considered as adsorption inhibitors. These are used for protection of wrapped components temporarily. The inhibitor is placed together with the component(s) and acts due to its suitable low saturation pressure, leading to a sufficiently durable inhibitor condensate on the metal surface. By this the effect of water and oxygen is prevented. It should be emphasized that these inhibitors may accelerate corrosion on some non-ferrous metals and alloys. Examples of vapour phase inhibitors are bicyclohexyl-ammonium nitrite and cyclohexyl-amine carbonate.

Inhibitors are very important for corrosion prevention in oil and gas production plants. The amounts, the cost, and the consequences of wrong treatment are so huge that selection and concentration of inhibitors must be decided by real specialists in the field. Companies specializing in inhibitors and other chemicals can be consulted. Oil companies have sponsored and carried out their own research on the subject, and may also hold a lot of knowledge about inhibitors themselves.

Use of inhibitors is generally most common in recirculation systems. For once-through flow systems the protection method is relatively expensive.

8.3.5 Proper Design

Design and materials selection are performed in connection with each other. In these processes both the individual components, the interactions between them and the relation to other structures and the surroundings have to be taken into account. The various phases of the life cycle of the construction, i.e. manufacturing, storing, transport, installation, operation and service, maintenance, and destruction should be considered.

The majority of the corrosion forms are affected more or less by the geometry: galvanic and thermogalvanic corrosion, crevice corrosion, erosion and cavitation corrosion, fretting corrosion, stress corrosion cracking and corrosion fatigue. In addition some general guidelines are useful:

(1) Design with sufficient corrosion allowance. Pipes, tanks, containers and other equipment are often made with a wall thickness twice the corrosion depth expected during the desired lifetime. However, this must of course depend on load and maximum permissible stresses too.

(2) Design such that the components that are most liable to corrosion are easy to replace. Special parts may be installed for "attracting" the corrosion. An example is shown in Fig. 8.21: a "spool" of carbon steel is inserted between a heat exchanger and the adjacent copper pipeline for the purpose that copper shall be deposited before the water enters the heat exchanger.

(3) For structures exposed to the atmosphere: the design should allow easy drainage with ample supply of air. Alternatively, the opposite: hinder air transport to cavities by complete sealing.

For components immersed in aqueous solutions there are similar extremes: efficient aeration should be secured (when this will cause passivation), or aeration should be prevented as far as possible.

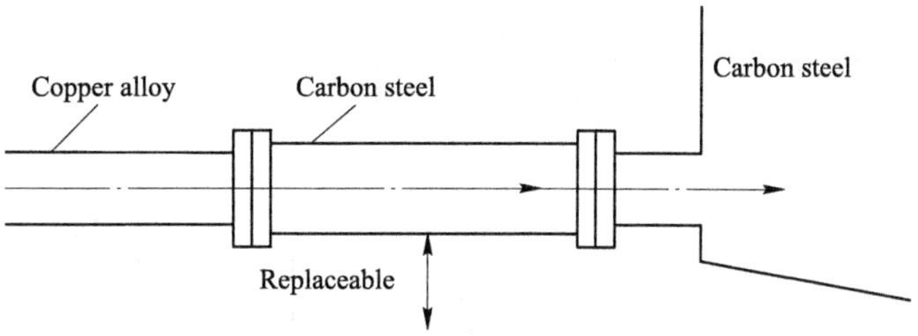

Fig. 8.21 The part with highest corrosion rate should be easy to replace

(4) Design in a way that makes drainage, inspection and cleaning easy (Fig. 8.22). Use joints that do not cause corrosion problems, e.g. butt welds instead of overlap joints. However, in cases

where the materials are protected by a paint or zinc coating before joining, sometimes it is beneficial to avoid welding. Alternatively, the welds can be post-treated (blast cleaned and coated, e.g. with paint, tape or thermal spray metal).

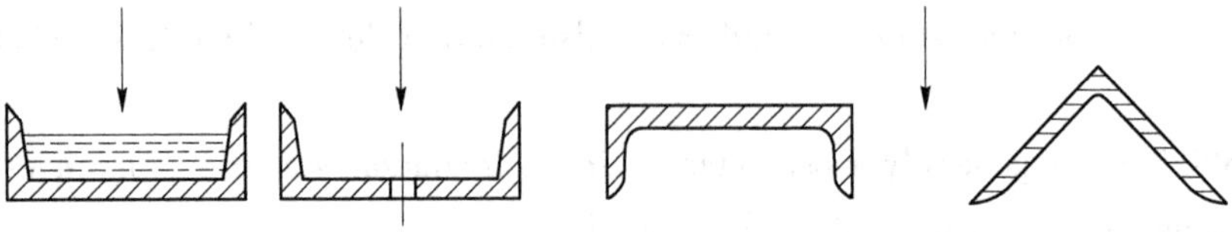

Fig. 8.22 Sections should be arranged in such a way that they can be drained and kept clean

(5) Avoid hot as well as cold spots. Heat exchangers and other equipment where heat transport occurs should be so designed that the surface temperature varies as little as possible. On superheated spots, increased (possibly thermogalvanic) corrosion will occur. For systems containing vapour, cold spots leading to local condensation and corrosion should be avoided (Fig. 8.23).

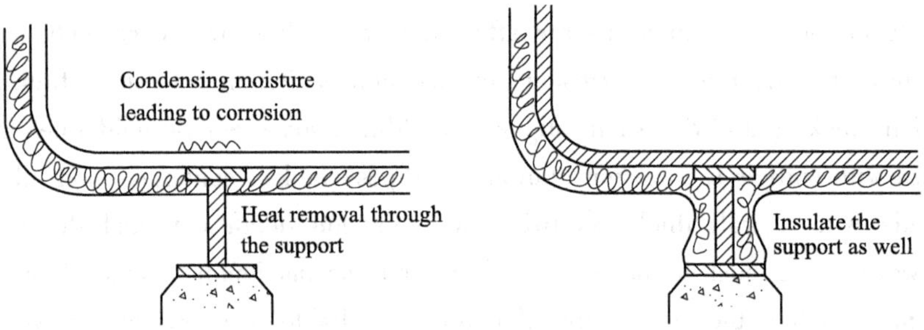

Fig. 8.23 Heat insulation of a vessel containing a moist gas

(6) Take the surroundings into account: make arrangements for minimizing the consequences of corrosion, e.g. where it may cause leakage (Fig. 8.24). Make sure that separate systems do not

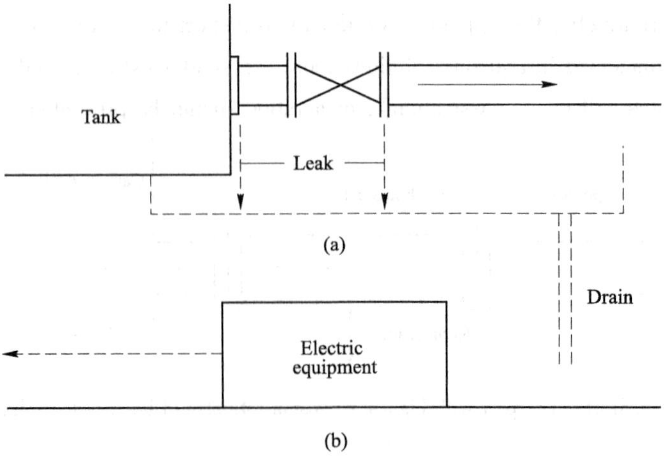

Fig. 8.24 Avoiding serious damage due to leakage

(a) Using a tray for collecting drips; (b) Moving sensitive equipment to a safer possition

impair the environment for each other. An example of this is if moisture containing copper ions (originating from corrosion of a copper alloy pipe) can come in contact withaluminium components.

(7) Avoid high corrosion risk on load-bearing parts or on critical places by shifting the attack to less critical places.

(8) Aim at simple geometry, and avoid heterogeneity and sharp changes in the system. Heterogeneity comprises different metals, uneven temperature and stress distribution, uneven dimensions etc. Abrupt dimension changes are unfortunate for stress distribution, temperature distribution, flow conditions, and possibly for the ease of cleaning. Sharp edges and irregularities on the surface impair the basis for painting, electrolytic plating and hot dipping (Fig. 8.25). Welds may contribute to several such problems. Crevices may cause harm in more than one sense, for instance on parts that are going to be pickled, since pickling acid may be entrapped and cause corrosion. Complex geometry and narrow gaps impede surface treatment such as blast cleaning, painting and thermal spraying, increase the cost, and make it more difficult to keep the structure clean and dry.

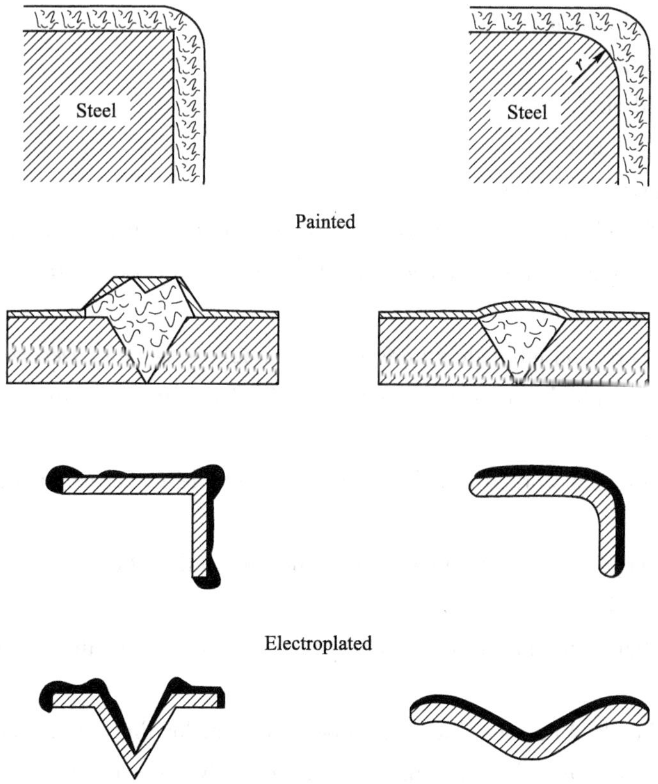

Fig. 8.25 Sharp edges and irregularities on the surface are unsuitable as a basis for coating

(9) Design for flow velocities that are compatible with the selected materials.

(10) Adjust pressure on gaskets (depending on gasket material) to avoid penetration of liquid

into crevices.

Questions

1. Tin cans are made of a steel, the inside of which is coated with a thin layer of tin. The tin protects the steel from corrosion by food products in the same manner as zinc protects steel from atmospheric corrosion. Briefly explain how this cathodic protection of tin cans is possible, given that tin is electrochemically less active than steel in the galvanic series.
2. From a molecular perspective, explain why increasing crosslinking and crystallinity of a polymeric material will enhance its resistance to swelling and dissolution. Would you expect crosslinking or crystallinity to have the greater in uence? Justify your choice.
3. List three differences between the corrosion of metals and each of the following: (a) The corrosion of ceramics; (b) The degradation of polymers.
4. Briefly explain the difference between oxidation and reduction electrochemical reactions. Which reaction occurs at the anode and which at the cathode?
5. For the following pairs of alloys that are coupled in seawater, predict the possibility of corrosion; if corrosion is probable, note which metal/alloy will corrode.
   (a) Aluminum and cast iron;
   (b) Inconel and nickel;
   (c) Cadmium and zinc;
   (d) Brass and titanium;
   (e) Low-carbon steel and copper.
6. Briefly describe the phenomenon of *passivity*. Name two common types of alloy that passivate.
7. Why does chromium in stainless steels make them more corrosion resistant than plain carbon steels in many environments?
8. For each form of corrosion other than uniform, do the following: (a) Describe why, where, and the conditions under which the corrosion occurs. (b) Cite three measures that may be taken to prevent or control it.
9. Briefly explain why cold-worked metals aremore susceptible to corrosion than non-cold worked metals.
10. Briefly explain why, for a small anode-to-cathode area ratio, the corrosion rate will be higher than for a large ratio.
11. For a concentration cell, briefly explain why corrosion occurs at the region having the lower concentration.
12. What are inhibitors? What possible mechanisms account for their effectiveness?
13. Briefly describe the two techniques that are used for galvanic protection.
14. Which of the following describes crevice corrosion? (a) Corrosion that occurs preferentially along grain boundaries; (b) Corrosion that results from the combined action of an applied tensile stress and a corrosive environment; (c) Localized corrosion that may be initiated at a surface defect; (d) Corrosion that is produced by a difference in concentration of ions or

dissolved gases in the electrolyte.

15. What properties are typical for a passivating oxide film? Mention examples of various types of surface films/surface layers.
16. A tank for storing an aqueous solution has a bottom made of stainless steel and a mantel made of unalloyed steel. The bottom and the mantel are welded together. The mantel and the weld are painted. Explain what happens if a small defect (a "holiday") arises in the paint coating on the unalloyed steel. How could this have been avoided? (There are various ways.)
17. For design of structures, equipment and single components it is, from a corrosion technology point of view, a good rule to aim at simple geometry, and to avoid heterogeneity and abrupt changes. Heterogeneity comprises different materials, varying temperature, uneven stress distribution and varying dimensions. Give examples of problems which can occur when this rule is neglected. As many corrosion forms and as many corrosion prevention methods as possible should be represented in the solution.

References

[1] Jones D A. Principles and prevention of corrosion[M]. New York: Macmilan, 1992.

[2] Costa J M. Progress in the understanding and prevention of corrosion: proceedings of the 10th European corrosion Congress[R]. 1993.

[3] Okubo M. Corrosion disaster and its prevention[J]. Journal of the Society of Materials Science Japan, 1978, 27:1027~1033.

[4] György Inzelt. Electrode potentials[M]. Barlin:Springer Berlin Heidelberg, 2013.

[5] Compton R G, Sanders G. Electrode potentials[M]. Oxfordshire: Oxford University Press, 1996.

[6] Arvidson E. Kristina Galvanic series of some dental alloys[J]. European Journal of Oral Sciences, 1977, 85(6):485~491.

[7] Gooch J W. Encyclopedic Dictionary of Polymers[M]. New York:spring, 2011.

[8] Dillon C P. Forms of corrosion: recognition and prevention[R] Forms of Corrosion Recognition & Prevention, 1982.

[9] Elayaperumal K, Raja V S. Forms of corrosion[R]. Hoboken: John Wiley & Sons, 2015.

[10] Perez N. Forms of corrosion[R]. Springer US, 2004.

[11] Unger L G. Certification of correction:US, US3010953 A[P]. 1961.

[12] Zhang X G. Galvanic corrosion[M]. Hoboken: John Wiley & Sons, 2011.

[13] Hack H P. Galvanic corrosion[J]. Reference Module in Materials Science and Materials Engineering, 2010.

[14] Rosenfeld I L, Marshakov I K. Mechanism of crevice corrosion[J]. Corrosion, 1964, 20(4):115~125.

[15] Syrett B C, Begum S. Corrosion, crevice[J]. Reference Module in Materials Science and Materials Engineering, 2016.

[16] Fontana M G. Corrosion engineering[M]. New York:McGraw-Hill Book Company, 1986.

[17] 木俣. Intergranular corrosion of ferritic stainless Steels[J]. Journal of the Japan Welding Society, 1955, 24.

[18] Mars G. Fontana New ideas on intergranular corrosion[J]. Industrial & Engineering Chemistry, 1950, 42(12):73A~74A.

[19] Hernandez S, Hassani S, Nassef A S. Erosion-corrosion[J]. Trends in Oil and Gas Corrosion Research and

Technologies, 2017:341~362.

[20] Postlethwaite J, Nešić S. Erosion-corrosion: recognition and control[J]. Uhlig's Corrosion Handbook, 2011: 907~913.

[21] Uhlig H H, Revie R W. Corrosion and corrosion control[M]. Hoboken: John Wiley & Sons, 1985.

[22] LaQue F L. Marine corrosion, causes and prevention[M]. Hoboken:John Wiley & Sons,1975.

[23] Osterheld C M. Corrosion prevention: US, US2490978 A [P]. 1949.

[24] Bardal E. Corrosion and protection[M]. London:Springer,2004.

[25] Tomashov N D, Chernova G P. Passivity and protection of metals against corrosion [M]. Boston: Springer, 2007.

[26] Lei W Y, Mi Y, Feng R J, et al. Hybrid 0D-2D black phosphorus quantum dots-graphitic carbon nitride nanosheets for efficient hydrogen evolution[J]. Nano Energy, 2018, 50: 552~561.